Springer Actuarial

MW00816564

Springer Actuarial Lecture Notes

This subseries of Springer Actuarial includes books with the character of lecture notes. Typically these are research monographs on new, cutting-edge developments in actuarial science; sometimes they may be a glimpse of a new field of research activity, or presentations of a new angle in a more classical field.

In the established tradition of Lecture Notes, the timeliness of a manuscript can be more important than its form, which may be informal, preliminary or tentative.

More information about this subseries at http://www.springer.com/series/15682

Ermanno Pitacco

ERM and QRM in Life Insurance

An Actuarial Primer

 Springer

Ermanno Pitacco
MIB Trieste School of Management
Trieste, Italy

Department DEAMS
University of Trieste
Trieste, Italy

ISSN 2523-3262   ISSN 2523-3270   (electronic)
Springer Actuarial
ISSN 2523-3289   ISSN 2523-3297   (electronic)
Springer Actuarial Lecture Notes
ISBN 978-3-030-49851-1   ISBN 978-3-030-49852-8   (eBook)
https://doi.org/10.1007/978-3-030-49852-8

Mathematics Subject Classification: G22, 97M30, 91B30

This Springer imprint is published by the registered company Springer Nature Switzerland AG
The registered company address is: Gewerbestrasse 11, 6330 Cham, Switzerland

Preface

The objective of this text is twofold. On the one hand, it aims to provide the reader with the basic concepts of Enterprise Risk Management in life insurance, with a specific focus on the quantitative phases of the Risk Management process, i.e. risk assessment, impact assessment and monitoring. On the other, the text stresses the need to extend the (traditional) actuarial toolkit in order to capture in quantitative terms the risk profile of life insurance lines of business.

The reader should then recognize a "bridge" between traditional actuarial mathematics, mainly based on the equivalence principle and hence deterministic, and risk-oriented approaches, based on stochastic models (but also on appropriate deterministic assessments) and allowing for diverse scenarios. It is also interesting to note that, while traditional actuarial mathematics relies on closed-form expressions, stochastic models adopted for risk and impact assessment call for computational approaches which usually resort to stochastic simulation.

The book is structured in nine chapters. After a brief Introduction (Chap. 1), in which relationships between risk management and actuarial work are presented, some definitions of Enterprise Risk Management (ERM) and Quantitative Risk Management (QRM) are provided and discussed (Chap. 2). The phases of the Risk Management (RM) process are then described (Chap. 3).

The remainder of the text focuses on ERM and QRM in the field of life insurance. First, the application of the RM process to life insurance and life annuity business is described (Chap. 4); special attention is placed on the "step" between the risk assessment phase and the impact assessment phase, as well as on RM actions aiming to face portfolio risks. Then, features of insurance products are addressed (Chap. 5), focusing on guarantees and options provided by some products, and risks consequently taken by the insurer.

Specific applications to life insurance (Chap. 6), life annuity (Chap. 7), and long-term care insurance business (Chap. 8) are then described to show the potential of an appropriate extension of the actuarial techniques. Several numerical results are also presented and discussed.

Finally, to provide a more complete picture of ERM in the insurance field, encompassing all the phases of the RM process, we propose a simple case study: the launch of a life annuity product (Chap. 9).

The final section of each chapter provides the reader with a list of references, which can help in extending or deepening the knowledge of the topics dealt with in the chapter itself.

The book has been planned and structured assuming as its target readers: advanced undergraduate and graduate students in Actuarial Sciences and in Insurance Economics and Finance, and professionals and technicians operating in life insurance, life annuity and pension areas.

It is assumed that the reader has attended courses providing the basics of actuarial mathematics and insurance techniques, in particular in the area of life insurance, as well as the basics of risk measures. Readers searching for such an introduction may wish to refer to the textbook by Olivieri and Pitacco (2015). The present text constitutes a natural addition to the latter, and is aimed at students wishing to pursue the actuarial education path.

The mathematics has been kept at a rather low level: indeed, all quantitative topics are presented in a "time-discrete" framework, and hence analytical tools like derivatives, integrals, etc. are not required.

The logical structure and the contents of these lecture notes have successfully been tested in various teaching initiatives; in particular:

- courses, short courses and lectures in Universities and Business Schools (Catholic University of Louvain, Louvain-La-Neuve; ESSEC Business School, Paris; European University at St. Petersburg; University of Trieste; MIB Trieste School of Management, Trieste);
- presentations in workshops and seminars (UNSW-CSIRO Workshop on "Risk: Modelling, Optimization and Inference", Sydney; BIRS-CMO Workshop on "Recent Advances in Actuarial Mathematics", Oaxaca, Mexico; seminars organized by the European Actuarial Academy, Munich, Stockholm, Madrid; AFIR-ERM seminars, Sofia and Florence; Institute of Actuaries of Japan, "120th Anniversary Special Seminar", Tokyo).

If this text helps the reader to understand the basic aspects of ERM and QRM in the life insurance and life annuity business, and stimulates the reader's interest in extending his/her knowledge of related topics, such as RM in health insurance or in pension funds, then it will have achieved its objective.

Trieste, Italy   Ermanno Pitacco
April 2020

Acknowledgments

The author would like to thank Dr. Daniela Y. Tabakova, researcher at the MIB Trieste School of Management, for her substantial help in planning and elaborating the numerical examples in Chaps. 6 and 7, as well as in reading the text and making comments. Of course, the author is responsible for any remaining errors and omissions.

The author would also like to thank the anonymous referees for their valuable comments and suggestions.

Contents

Chapter 1
Introduction

Risk Management (RM) constitutes a very recent achievement for the actuarial culture and profession, especially if compared to the long history of actuarial mathematics. Why so recent? An appropriate answer calls for some remarks about traditional features of the actuarial sciences.

We first note that, for a long time, the only traditional target of actuarial education and training were the insurance activities. In particular, the focus was initially restricted to life insurance and pensions, later extended to non-life insurance, and even later to financial issues. The insurance activities involve, by their nature, the management of *risks*, but the original actuarial setting disregarded, for a long period, a rigorous risk-oriented approach, especially in life insurance. Indeed, the equivalence principle, only based on expected values and widely adopted in the life insurance context for pricing and reserving purposes, does not explicitly account for risks, whose impact can only implicitly be faced, at the pricing level, via an implicit safety loading.

Conversely, the presence of risk awareness is clearly witnessed by the underwriting process (aimed at assessing the risk at a policy level), the risk transfer via reinsurance (at policy level or at portfolio level), and the capital allocation.

However, the strength of many traditional actuarial ideas (e.g., the equivalence principle in life insurance, the risk theory principles in non-life insurance, the experience-rating mechanisms, etc.) contributed to assign an exclusive role to classical actuarial techniques in solving insurance quantitative problems. As a side-effect, we can recognize a historical "closure" of the actuarial techniques in respect of other disciplines.[1]

New scenarios clearly emerged in the last decades of the twentieth century, which involved financial aspects, biometric issues, legislation and regulation, market competition. New insurance products were designed and launched, especially in the area of the insurances of the person (life and health insurance in particular).

---

[1] For an extensive review of the early contributions to actuarial mathematics and insurance techniques, the reader can refer to Haberman (1996).

© Springer Nature Switzerland AG 2020
E. Pitacco, *ERM and QRM in Life Insurance*, Springer Actuarial,
https://doi.org/10.1007/978-3-030-49852-8_1

Traditional actuarial ideas and techniques then revealed a serious lack of power in dealing with all the new issues originated by the scenario evolution. Focusing on life insurance, some examples are given by: trends in longevity and related uncertainty (the "aggregate" longevity risk), decreasing interest rates, new European principles in the calculation of mathematical reserves and (also in non-life insurance) a new solvency regime and related capital requirements.

Hence, contributions from other disciplines (e.g., finance, corporate finance, accountancy, etc.) were required. However, importing ideas and methods into the insurance field, traditionally relying on actuarial methods and techniques, originated a rather chaotic situation regarding concepts, terminology, etc. Conversely, a sound risk-oriented approach to management (and to insurance management, in particular) contributed to re-organize the insurance activity, to innovate the insurance disciplines, and to improve the related "language". And, regarding the language:

Die Grenzen meiner Sprache sind die Grenzen meiner Welt,
Ludwig J. J. Wittgenstein, *Tractatus Logico-Philosophicus* (prop. 5.6), 1921

RM principles entered into the insurance field, without "destroying" the actuarial methods. On the contrary, actuarial methods and relevant applications have been extended in order to properly capture risk-related issues.

In the innovation process, a (possible) short step could be the following one: from *RM in insurance* to *actuaries in RM*. A strong point, which can ease this step, can be found in the probability-based background of the actuarial profession, i.e. a background needed for properly dealing with stochastic models for risk assessment. Conversely, a (possible) weak point can be the lack of a "corporate" perspective and of management skills in the (traditional) actuarial education.

However "reciprocal" contributions can be extremely profitable, especially (but not only) in the insurance and finance field.

On the one hand, Enterprise Risk Management (ERM, see Chap. 2) provides guidelines for a sound management of risks, with a wide range of applications: banking, insurance, commerce, as well as other sectors of production. Of course, guidelines are general purpose, hence their implementation is, to a large extent, sector-specific (or even entity-specific).

On the other hand, actuaries can significantly contribute to the implementation of RM "phases", including (but not limited to) the Quantitative Risk Management (QRM) (see Chap. 3), in particular:

- risk assessment and impact assessment, which call for the design and the implementation of appropriate stochastic models;
- monitoring, which also relies on the statistical analysis of past experience.

At the same time, ERM should improve actuaries' awareness of corporate issues, for example: capital allocation, creation of value, organization, etc. In particular, in the insurance context, ERM can suggest to actuaries solutions to new problems as well as new solutions to old problems.

Stochastic approaches and relevant implementations must play a prominent role in ERM, and specifically in QRM. Nevertheless, the power of appropriate deterministic approaches should not be underestimated. Deterministic assessments can indeed help the actuary in achieving a comprehensive view of all the possible impacts of risks on results of interest. In particular, they can provide a first insight into ranges of results, as well as they can point out the impact of extreme scenarios. The desirable methodological features of the actuarial work could then be summarized by the following invitation: "be stochastic but don't forget to be, from time to time, reasonably deterministic".

Chapter 2
Enterprise Risk Management (ERM) and Quantitative Risk Management (QRM)

2.1 What Is ERM?

A number of definitions of Enterprise Risk Management (ERM) have been proposed. We recall some of them, cited in the report IAA (2009).[1]

2.1.1 Some Definitions

The following definition has been proposed by COSO (2004)[2]:

> ERM is a process, effected by an entity's board of directors, management and other personnel, applied in strategy setting and across the enterprise, designed to identify potential events that may affect the entity, and manage risk to be within its risk appetite, to provide reasonable assurance regarding the achievement of entity objectives.

The definition proposed by CAS (2003)[3] is as follows:

> ERM is the discipline by which an organisation in any industry assesses, controls, exploits, finances, and monitors risks from all sources for the purpose of increasing the organisation's short- and long-term value to its stakeholders.

KPMG (2001) proposed the following definition:

> ERM is a structured and disciplined approach aligning strategy, processes, people, technology, and knowledge with the purpose of evaluating and managing the uncertainties the enterprise faces as it creates value.

Another definition proposed by KPMG in 2005 is as follows:

---

[1]IAA = International Actuarial Association (https://www.actuaries.org/iaa).

[2]COSO = Committee of Sponsoring Organizations of the Treadway Commission, US (https://www.coso.org/Pages/default.aspx).

[3]CAS = Casualty Actuarial Society, US (https://www.casact.org/).

© Springer Nature Switzerland AG 2020
E. Pitacco, *ERM and QRM in Life Insurance*, Springer Actuarial,
https://doi.org/10.1007/978-3-030-49852-8_2

ERM is the process of planning, organising, leading, and controlling the activities of an organisation to minimise the effects of risk on an organisation's capital and earnings.

The IIA (2009)[4] provided the following definition:

ERM is a structured, consistent and continuous process across the whole organization for identifying, assessing, deciding on responses to and reporting on opportunities and threats that affect the achievement of its objectives.

In IAA (2009) we find the following definition:

Enterprise Risk Management, as described here, is a holistic management process applicable in all kinds of organisations at all levels and to individuals. ERM differs from a more restricted "risk management" used in some sectors. For example, in some areas the terms "risk management" or "risk control" are used to describe ways of dealing with identified risks, for which we use the term "risk treatment". Some other terms used in this document also have different usages. For example the terms "risk analysis", "risk assessment" and "risk evaluation" are variously used in risk management literature. They often have overlapping and sometimes interchangeable definitions, and they sometimes include the risk identification step.

Moving to the specific insurance context, we refer to the Insurance Core Principles (ICP) by IAIS (2019),[5] which focuses on ERM in insurance companies for solvency purposes. In ICP 16 we find, in particular, what follows:

ERM for solvency purposes is the coordination of risk management, strategic planning, capital adequacy, and financial efficiency in order to enhance sound operation of the insurer and ensure the adequate protection of policyholders. Capital adequacy measures the insurer's assessment of residual risk of its business, after overlaying the mitigating financial effect of the insurer's established risk management system. Any decision affecting risk management, strategic planning or capital would likely necessitate a compensating change in one or both of the other two. Successful implementation of ERM for solvency purposes results in enhanced insight into an insurer's risk profile and solvency position that promotes an insurer's risk culture, earnings stability, sustained profitability, and long-term viability, as well as the insurer's ability to meet obligations to policyholders. Collectively practiced in the industry, ERM for solvency purposes supports the operation and financial condition of the insurance sector. These aspects of ERM should therefore be encouraged from a prudential standpoint.

The ERM framework for solvency purposes (ERM framework) is an integrated set of strategies, policies and processes, established by the insurer for an effective implementation of ERM for solvency purposes.

2.1.2   Some Ideas

The use of the term *holistic* emphasizes the importance of the "whole" and the interdependence of the "parts". Indeed, ERM must account for *risks of all sources*, and, to this purpose, calls for a perspective which looks *across the enterprise*.

---

[4]IIA = Institute of Internal Auditors (https://global.theiia.org/Pages/globaliiaHome.aspx).

[5]IAIS = International Association of Insurance Supervisors (https://www.iaisweb.org/home).

Table 2.1 A paradigm shift—*Source* Gorvett (2006)

Traditional	Emerging
Risks managed "in silos"	Centralized management with executive-level coordination
Concentrates on physical hazards and financial risks	Integrated consideration of all risks, firm-wide
Insurance orientation	Opportunities for hedging, diversification
Ad-hoc/one-off projects	Continuous and embedded

The term "risk" is often used, especially in the common language, with a "negative" meaning. According to an appropriate language, we have to distinguish between *pure risks* (also named *non-value-adding risks*), which can only lead to losses (e.g., the loss caused by a fire), and *speculative risks* (or *value-adding risks*), which can either lead (hopefully) to profits or to losses, e.g., the result of a financial transaction. Whatever the type of risk might be, the uncertainty concerning specific events originates the randomness of the profit or loss.

In the ERM context, the expression *risk appetite* is commonly used to denote the acceptance of possible losses which may result from the aggregation of all the risk sources, speculative risks of course included. Sometimes, the expression *risk tolerance* is conversely used to denote the accepted level of loss originated by each risk source.

2.1.3 Some Aspects of Evolution

Approaches to the management of risks involved by an activity have significantly evolved throughout time. Important features of this evolution have been singled out by Gorvett (2006). Tables 2.1 and 2.2 summarize the main features. We note that, in Table 2.2, two documents[6] are referred to, which can be considered milestones in the ERM evolution.

---

[6] The *Turnbull Report* (short name for "Internal Control: Guidance for Directors on the Combined Code"), issued in 1999, was drawn up with the London Stock Exchange for listed companies, to inform directors of their obligations with regard to keeping good internal controls in their companies and checks to ensure the quality of financial reporting.

The *Sarbanes-Oxley Act* is a US Federal law, officialy known as "Public Company Accounting Reform and Investor Protection Act", which was enacted in 2002 as a reaction to a number of major corporate and accounting scandals. These scandals lead to investors' losses of billions of dollars when the share prices of affected companies collapsed. The Act created a new, quasi-public agency, charged with overseeing, regulating, inspecting, and disciplining accounting firms in their roles as auditors of public companies.

Table 2.2 Evolution of ERM—*Source* Gorvett (2006)

Period	Approach
Historically	"Risk silo" mentality
Mid-1990s	First "Chief Risk Officer"
	First use of ERM terminology
Late-1990s	Risk-related regulatory requirements (e.g., the Turnbull Report; UK, 1999)
2001–2002	September 11: new awareness on risk correlation
	Corporate scandals (Enron, Arthur Andersen, Tyco International)
	Efforts to improve corporate governance: Sarbanes-Oxley Act in the US

2.1.4   Some Terms (and the Underlying Ideas)

In the ERM language, expression such as *assessing, evaluating ... risks* are frequently used. A quantitative area of ERM is then involved, and this constitutes a typical example of actuarial intervention in the RM process.

More specifically, when referring to *risks* and *effects* or *impacts of risks* a three-phase approach is involved:

- identifying the risks, i.e. the random variables (e.g., in the insurance field: return on investments, inflation rates, company market share, number of claims in an insurance portfolio, etc.), which can impact on results;
- quantifying the risks via appropriate stochastic models;
- quantifying the relevant impacts, i.e. the effects (e.g., on cash flows, assets, net asset value, etc.) which are functions of the risks, and looking for the relevant probability distributions.

2.2   What Is QRM?

Quantifying the risks and the relevant impacts (see Sect. 2.1.4) constitutes the core of the QRM. Implementing QRM calls for a set of mathematical models and calculation tools.

2.2.1   The QRM Toolkit

A list of models and tools for implementing QRM follows.

1. General probabilistic models: probability distributions of random variables, copulas to express correlations among random variables, etc.
2. Specific probabilistic models, in particular for:

    a. financial applications (e.g., CAPM, geometric Brownian motion, Black & Scholes model, etc.);
    b. analysis of distribution tails, to quantify "extreme" risks.

3. Synthetic values, in particular risk measures (e.g., VaR, TailVaR, etc.).
4. Calculation procedures are needed to obtain (usually approximate) probability distributions of impacts of risks; in particular:

    a. analytical methods, possibly based on closed-form expressions (however, in most applications: mission impossible!);
    b. recursive calculations (e.g., algorithms for determining loss distributions in insurance);
    c. stochastic (or Monte Carlo) simulation, possibly based on scenario generators, providing "empirical" distributions of the impact.

5. A simpler assessment of impacts of risks can be achieved via "sensitivity" testing, that is, by assigning alternative values to the random variables which represent risks (e.g., return on investments, inflation rate, etc.) or by assuming diverse scenario hypotheses (e.g., mortality, disability, etc.), and then calculating the corresponding values of quantities of interest (cash flows, net asset value, etc.).

Calculation procedures in life insurance and life annuities, which aim at obtaining probability distributions (and hence relying on stochastic simulation, see point 4c above), will specifically be addressed in Chaps. 6 and 7. Procedures aiming at performing a sensitivity testing (see point 5) will also be addressed in Chaps. 6 and 7 as well as in Chap. 8.

QRM is, within the ERM framework, the typical (but not the only) field for actuarial work: indeed, a sound mathematical and probabilistic background is needed. Actuaries (more than other mathematical skilled professionals) should have appropriate skills, in particular related to finance, corporate finance, financial reporting.

2.2.2 QRM: Solvency and Internal Models

Solvency requirements (according to Solvency II, regarding insurance activities) are determined, for each entity, relying on:

1. standard formulae;
2. internal models;
3. a mix of 1 and 2, consisting of "partial" internal models, that is, models which focus on:

   a. the impact of risks on a specific set of line of business (LOBs), e.g., specific portfolios of an insurance company;

   b. the impact caused by a specific set of risk causes, e.g., return on investments and stock market indexes, on the results of all the LOBs involved.

The choice among 1, 2 and 3 is in particular constrained by the availability of human resources and computational budgets. If approach 2 or 3 is chosen, then QRM is clearly involved.

Whatever the choice, ERM is involved as a "mentality". As mentioned in Sect. 2.1, basic features are:

- integrated consideration of all risks, that is, "firm-wide";
- centralized management with executive-level coordination.

2.3 References and Suggestions for Further Reading

In this section, we first cite textbooks dealing with general aspects of RM, with special focus on ERM and QRM. Studies particularly devoted to applications in life insurance will be cited in the relevant sections of the following chapters.

ERM is addressed by CAS (2003), Chapman (2006), COSO (2004), Dickinson (2001), Lam (2003), Olson and Wu (2008), and Rochette (2009). The book by Sweeting (2017) specifically focuses on financial ERM.

The Risk Book by IAA (n.d.) provides a significant contribution to the risk culture. The IAA Risk Book is a collection of papers which constitute high quality reference material in the field of RM. Aims of the book are presented in its Introduction by Sandberg (2015). Several papers will be cited in the following chapters of the present book, according to their specific contents.

While value creation constitutes one of the objectives of any company RM, implementation of appropriate ERM procedures can significantly contribute to the value creation and to the rating of the company itself. The reader is referred, for example, to the report by KPMG (2001). Among the most recent contributions, see: Bohnert et al. (2017, 2019), Farrell and Gallagher (2015), and Hoyt and Liebenberg (2015).

Gorvett (2006) deals with the role of auditing in ERM, while IIA (2009) focuses on internal auditing in ERM.

A mathematical framework, within which ERM can be viewed and all the concepts of ERM can be interpreted, is proposed by Taylor (2013).

The books by Doherty (2000), Harrington and Niehaus (1999), and Williams et al. (1998) offer comprehensive presentations of the RM process (which will be addressed in Chap. 3), the insurance transfers included. Practical guidelines to RM in business and industry are provided by Koller (1999).

The process of analyzing and planning for both personal risks and business risks is examined by Rejda (2010). The books by Mangiero (2005) and Crouhy et al. (2001) focus on RM in financial institutions.

The Insurance Core Principle 16 in IAIS (2019) stresses the importance of an appropriate ERM framework in insurance companies for solvency purposes.

Moving to quantitative issues, we first cite the book by McNeil et al. (2005), which provides a comprehensive presentation of the basic ideas underlying QRM, as well as a number of models and technical tools.

Jorion (2007) is the classical reference on Value at Risk (VaR). The book by Dowd (1998) deals with RM and the VaR approach to RM problems, whereas Pearson (2002) mainly focuses on the use of VaR in portfolio management.

Klugman et al. (2012) deals with the construction of probabilistic models quantifying losses, suitable in decision processes, starting from available data.

Lists of books, papers, articles, technical reports on ERM-related topics are maintained by actuarial associations and universities. In particular, we recommend the website ERM Library maintained by the Society of Actuaries in Ireland (https://web.actuaries.ie/press/erm-library), which provides a collection of ERM-related scientific and practical material. We also recall the website of the North Carolina State University—Poole College of Management—Enterprise Risk Management Initiative (https://erm.ncsu.edu/library) where, besides papers and reports on case-studies, also recorded presentations, roundtable summaries and discussions on ERM-related topics are available.

Chapter 3
The Risk Management (RM) Process

Given the introductory aims of this chapter, concepts and examples are referred, in some cases, to insurance business, in other cases to non-insurance activities. Conversely, Chaps. 4–9 will specifically refer to life insurance and life annuity business.

3.1 Phases of the RM Process: An Overall View

The implementation of RM principles takes place via the *risk management (RM) process*, which basically consists of a sequence of phases, as sketched in Fig. 3.1.

3.1.1 Setting the Objectives

Any organization aims at achieving given objectives. Important objectives are listed below.

- *Profit*. According to an accounting perspective, profit can be defined as the difference between revenues and costs, that is, the "accounting earnings". While reducing the negative impact of risks on various quantities (e.g., cash flows, assets, etc.) is an obvious target of the RM process (see "risk mitigation"), a reasonable amount of risk appetite is needed in any organization which aims at an appropriate potential for profit.
- *Value creation*. A number of different meanings can be attributed to the word "value" and hence to the expression "value creation", also depending on the context and the stakeholders referred to, e.g., the clients, the shareholders, etc. When referring to the shareholders, value creation can be meant as a synonym for (positive) "economic earnings". Economic earnings can be defined as the difference

© Springer Nature Switzerland AG 2020

E. Pitacco, *ERM and QRM in Life Insurance*, Springer Actuarial,
https://doi.org/10.1007/978-3-030-49852-8_3

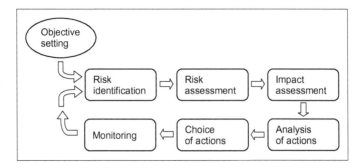

Fig. 3.1 Phases of the RM process

between the revenues and the costs associated to all of the production factors, hence including the cost of shareholders' capital invested in the business.

- *Market share/New business.* Keeping and possibly increasing the market share is an obvious target for all the organizations which sell products. Creating value for the clients can improve the market share.
- *Rating.* Organizations receive their financial strength rating from rating companies which look at financial performance both past, present and future. High ratings have obvious advantages, for example: opportunity to reduce the cost of borrowing, improvement of the organization image, ability of attracting investors, etc.
- *Risk mitigation.* Risk mitigation aims at reducing the expected total impact (loss) of risks. In particular, risk mitigation actions tend to lower the expected number of occurrences (*risk prevention*), or to reduce the expected amount of each loss (*risk reduction*), or both.
- *Solvency.* Commonly, the term "solvency" is used to denote the capability of an organization to pay all the amounts when these fall due. More specific definitions of solvency (in probabilistic terms) are needed when referring to insurance activities.
- *Loss financing.* Although negative impacts of risks can be limited by risk mitigation, losses may occur and must then be met via appropriate financing solutions. Thus, an important objective is the access to internal or external resources in case of need.

3.1.2   From Risk Identification to Monitoring

To achieve the objectives selected by the organization, the following RM phases must be implemented.

1. *Risk identification.* In this phase the risk causes, that is, the causes of potential losses suffered by the organization (the business, or the family, or the individual), as well as the profit sources, are singled out.

2. *Risk assessment.* Risk causes are expressed in quantitative terms via appropriate stochastic models (viz probability distributions).
3. *Impact assessment.* The impact of risk causes on results of interest (cash flows, assets, profits, value creation, etc.) is quantified in terms of probability distributions of the results themselves, and related typical values (expected values, variances, VaR, TailVaR, etc.).
4. *Analysis of actions.* Available RM actions are listed (pricing of the products, risk transfer via insurance, capital allocation, etc.), and costs and benefits related to available actions are compared.
5. *Choice of actions.* Usually an appropriate mix of actions is chosen (e.g., combining risk transfer and capital allocation).
6. *Monitoring.* This phase should involve both the results achieved by the organization and the assumptions about the scenario (e.g., behavior of the capital markets, inflation rate, tax legislation, etc.) adopted when choosing RM actions.

From Fig. 3.1, the never-ending characteristic of the RM process clearly emerges.

3.2   From Data to Models, from Models to Results

While risk assessment and impact assessment are the core of QRM, as mentioned in Sect. 2.2, risk identification obviously constitutes the preliminary phase. It aims to single out the potential loss exposures of an organization. Causes of risks concerned clearly depend on the particular organization under analysis. All these phases must be supported by data. What data are relevant to the specific problem is a matter of judgement. Then, the role of the "expert" clearly appears. See Fig. 3.2.

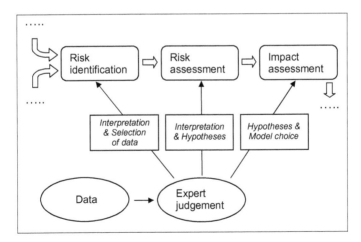

Fig. 3.2 Expert judgement in QRM

3.2.1   Data and Results

Expert judgement must underpin, via appropriate interpretation, the choice of data needed in the risk identification phase. More data make the following assessment phases more complex but closer to the reality, provided that good quality data are available. Disregarding data simplifies the assessments, but might introduce biases.

Referring to a life insurance portfolio, while investments yields, insureds' mortality, lapses and surrenders must be included into the analysis, the expert may wonder about the use of data like the trend in average personal income as explaining the propensity to buy insurance and hence the possible future market share of the insurance company.

In the risk assessment phase, each cause of risk must be expressed in quantitative terms. In the RM practice, frequently synthetic values, e.g., frequencies of relevant events, are only considered. A more accurate (and complex) approach to risk and impact assessment should involve the use of appropriate probabilistic models. For example, again referring to a life insurance portfolio, life tables must be chosen to quantify the insureds' age-pattern of mortality. Hence, average numbers of people dying in the various years can be easily calculated. However, this way no information about the variability of the numbers of deaths can be achieved. To this purpose, a probability distribution of the random number of deaths is required, for example a Binomial or a Poisson distribution. Specific hypotheses must then be adopted, according to expert judgement. It should be stressed that all the choices involved in the risk assessment phase constitute a bottleneck with regard to the impact assessment phase: if risk is only assessed in terms of synthetic values, just synthetic values will summarize the impacts.

The purpose of the impact assessment phase is to quantify in monetary terms the consequences of the risks borne by the organization. Quantities such as cash flows, profits, etc. should be addressed. Moving from risk assessment to impact assessment can be a challenging step when complex realities are concerned. Indeed, probability distributions related to risk causes (the input) must be "transformed" into probability distributions describing impacts (the output). The shift from risk assessment to impact assessment calls for appropriare hypotheses, also looking for tractability of the resulting model. This can be, for example, the case of a life insurance or life annuity portfolio, as we will see in Chaps. 6 and 7. Some typical values (expected value, variance, mode, and so on) should anyway be focused on, as these can help in comparing various situations and then choosing actions.

3.2.2   Modeling and Communication

Risk identification and, more specifically, risk assessment and impact assessment phases result in the construction of a *model*, which links results of interest to the various risk causes. Modeling issues, with reference to life insurance, will be dealt

Fig. 3.3 QRM phases:
interactions among processes
in an insurance company

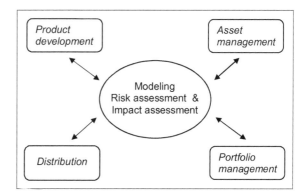

with in Sect. 4.4. Now we only stress that the process leading to the construction of the model must rely, according to the ERM logic, on appropriate interactions among the various "processes" and "functions" in the organization. An example is sketched in Fig. 3.3, again with reference to life insurance.

Remark Every model is a simplified representation of the reality and, because of the simplifications, cannot of course exactly reproduce the reality. As noted by George E. P. Box in 1978, "All models are wrong, but some are useful". In 1987, Box then stressed that: "The practical question is how wrong do they have to be to not be useful". In other words, while approximations can (and, frequently, must) be accepted, biases should carefully be avoided.

Effective interactions call for appropriate communication among company functions and processes. While detailed and analytical information about results of interest is technically needed for choosing RM actions, summaries and graphical representations can help in providing a general picture of the organization risk profile.

We focus on a tool recently proposed in the ERM framework. A *risk heat map* is a two-dimension graphical tool used to present the results of the risk assessment and impact assessment phases. The impact of each risk cause is represented in terms of expected frequency and severity (the average amount of loss if a relevant event occurs). Quantification of frequency and severity results from the two phases of the RM process.

An example is given by Fig. 3.4. Moving from green boxes to red boxes, both frequency and severity increase: thus, the expected impact, simply expressed by the product frequency × severity, increases. For example, in Fig. 3.4 risk 1 has a very low impact, whereas risk 9 has a very high impact. Risks 4, 5, 6 and 7 have a medium impact, but the impact itself is the result, for each risk, of a different combination of frequency and severity. The diversity of the two factors suggests different RM actions, as we will see in Sect. 3.5.1.

An effective communication tool is given by the *Key Risk Indicators* (*KRIs*) which provide information on emerging risks, hence working as "early warning indicators". KRIs can refer to the organization as a whole, or to a specific line of business.

Fig. 3.4 Risk heat map

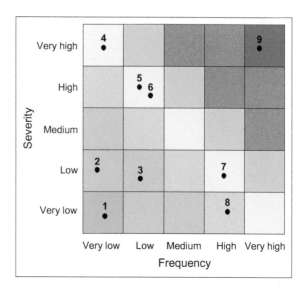

Emerging risks can better be captured looking at past trends in some results of interest, and summarizing the trends, for example in terms of average annual variations in the results. In the life insurance area, increasing trends in lapses can constitute a significant emerging risk. The trends can be expressed by appropriate KRIs, either in terms of ratios of annual number of lapses to number of policies in force, or in terms of ratios of total amounts assured in policies lapsed to total amounts assured.

3.3 Looking at Insurance Companies: ERM and ORSA

The scope of insurance regulations (e.g., Solvency II and, in particular, its second pillar) extends beyond the quantitative capital requirements, encompassing "qualitative" requirements. A significant quality requirement is the set-up and development of an effective risk management framework, as suggested by ERM principles. In this context, *Own Risk and Solvency Assessment (ORSA)* plays a substantial role which results, in particular, in information communicated to regulators.

However, ORSA should not only be meant as a regulation requirement, but as a collection of "internal" processes including (but not limited to) internal control, internal audit, and, of course, the RM process. The RM process, as defined in Sect. 3.1, is the major tool for risk identification, risk and impact assessment, and analysis and choice of actions. As already noted, the quantitative phases of the RM process (that is, the QRM) provide a substantial help in choosing appropriate RM actions.

The logical "ordering by inclusion" of the above concepts is sketched in Fig. 3.5. However, it should be stressed that interactions among processes have to work inside

Fig. 3.5 ERM, ORSA and
the RM process

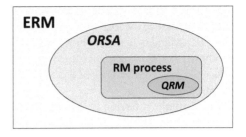

the ORSA and the broader ERM framework. An example, referred to the construction
of a QRM model in insurance, is provided by Fig. 3.3.

3.4 RM: Objectives and Actions

RM actions can be classified according to various criteria. In what follows, we focus
on some RM actions, grouped according to objectives (risk mitigation and loss financing, in particular). Special attention is then placed on risk hedging.

Specific objectives and RM actions in the field of life insurance and life annuities
will be addressed in the following chapters.

3.4.1 General Issues

The *analysis of actions* aims at singling out what RM actions are available to achieve
the organization objectives, and in particular to face risks and relevant impacts, and
to compare costs and benefits related to each action.

The RM actions can be classified according to their objective as displayed in
Table 3.1.

Table 3.1 Classification of RM actions

Objective	Actions
Risk mitigation	Risk prevention
	Risk reduction
	Risk avoidance
Loss financing	Retention
	Risk transfer via insurance
	Other contractual risk transfers
Internal risk reduction	Diversification
	Investment in information

Actions aiming at *risk mitigation* (also called *risk control*, or *loss control*) generally tend to reduce the total expected loss due to some events in a given period (e.g., a year). In particular, actions which tend to lower the expected number of occurrences are known as *risk prevention* (or *loss prevention*) actions, whereas actions aiming to reduce the expected severity of each loss are called *risk reduction* (or *loss reduction*) actions.

For example, referring to the risk of fire in a factory, appropriate electric equipments can contribute in reducing the expected number of fire occurrences (risk prevention), whereas fire protection measures (e.g., appropriate doors) can lower the risk of fire propagation and hence the expected amount of damage (risk reduction).

Risk control can also be realized by reducing the level of risky activities, in particular by shifting to less risky product lines. Clearly, the cost of this action is given by a reduction in the profits produced by the risky activities. The limit case is given by the total elimination of these activities: this action thus aims at *risk avoidance*.

The expression *loss financing* (sometimes *risk financing*) denotes a wide range of methods which aim at obtaining financial resources to cover possible losses, anyhow unavoidable.

First, the organization can choose the *retention* of the obligation to pay losses. Retention is often called *self-insurance*. Instead of retaining a risk, the organization can transfer it to another organization. The usual transfer consists in the *insurance* of the risk, and thus involves, as the counterpart, an insurance company. Nevertheless, other risk transfer arrangements can be conceived. More details on this topic can be found in Sect. 3.5.

Finally, we turn to actions aiming at *internal risk reduction*. *Diversification* typically relates to investment strategies and related risks, and consists in investing relatively small amounts of wealth in a number of different stocks, rather than putting all of the wealth into one stock. Diversification makes the investment results not totally depending on the economic results of just one company, and hence aims at the reduction of the investment risk.

Investment in information is the second major action of internal risk reduction. Appropriate investments can improve the "quality" of data (see Sect. 3.2.1), e.g., estimates and forecasts. A reduced variability around expected values follows, so that more accurate actions of, for example, loss financing can be performed.

The analysis of alternative actions must be followed by the *choice of actions* to be implemented. Actions in RM are not mutually exclusive, so that the strategy actually adopted usually consists of an appropriate mix of several actions. For example, risk prevention and risk reduction can be accompanied by an appropriate insurance transfer, which, in its turn, will be less expensive if an effective risk mitigation can be proved to the insurer.

3.4.2 Hedging

In the ERM language, the expression "*hedging* strategies" is usually attributed a broad meaning, which encompasses diverse actions aiming at risk mitigation, and possibly loss financing.

In particular, hedging can be based on the use of financial derivatives, such as futures, forwards, swaps, options, and so on. These derivatives can be used to offset potential losses caused by changes in commodity prices, interest rates, currency exchange rates, and so on. For example, a factory which uses oil in the production process is exposed to losses due to unanticipated increases in the oil price. This risk can be hedged by entering into a forward contract, according to which the oil vendor must provide the user with a specified quantity of oil on a specified date at a price stated in the contract.

Hedging strategies can be described in rather general terms, so that a number of RM actions, risk transfers in particular, can be placed in the hedging framework.

Consider a stream of random outflows X_t, $t = 1, 2, \ldots$ (costs, expenses, losses, etc.). Let $X_t \geq 0$, and assume, for simplicity, that all the X_t are identically distributed. Their outcomes fluctuate over time. A reduction in the variability is our objective.

Basic ideas are sketched in Table 3.2. High outcomes of X_t are denoted by the event $[X_t = \text{high}]$, whereas low outcomes are denoted by $[X_t = \text{low}]$. Hedging flow streams can be implemented, such that combining the original outflow stream X_t, $t = 1, 2, \ldots$, with an appropriate hedging flow stream results in a flow stream with a lower variability. The features of the hedging flows have to be in line with a specific target. Two basic approaches can be adopted to define a hedging strategy.

1. Offset the random outflow X_t with a hedging random flow, such that the variability of the sum of the two flows is smaller than the variability of X_t. This target can be achieved as follows:

   (a) the sign of the hedging flow Y_t is opposite to the sign of X_t (and hence Y_t is an inflow), and is larger (in absolute value) the larger is X_t; small values of X_t imply small values of Y_t, possibly $Y_t = 0$; hence, reduction in variability is achieved by obtaining a net outflow $X_t + Y_t$ smaller than (or equal to) the original outflow X_t, and in particular not greater than a stated upper bound (*cap*) x';

Table 3.2 Hedging strategies

Outflow X_t	$[X_t = \text{high}]$	$[X_t = \text{low}]$
1. Hedging flow		
(a) Inflow Y_t	$[Y_t = \text{high}]$	$[Y_t = \text{low}]$
(b) Outflow Z_t	$[Z_t = \text{low}]$	$[Z_t = \text{high}]$
2. Hedging swap		
Inflow/Outflow W_t	$[W_t < 0]$	$[W_t > 0]$

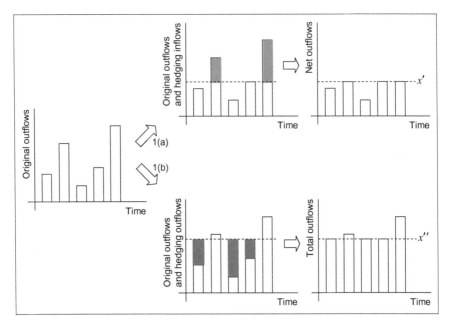

Fig. 3.6 Hedging strategies (1)

(b) the hedging flow Z_t has the same sign of X_t (and hence Z_t is an outflow), and is smaller the larger is X_t; large values of X_t imply small values of Z_t, possibly $Z_t = 0$; reduction in variability is then achieved by constructing a total outflow $X_t + Z_t$ larger than (or equal to) the original outflow X_t, and in particular not smaller than a stated lower bound (*floor*) x''.

2. Choose an upper bound $x^{(1)}$ and a lower bound $x^{(2)}$. Let W_t denote the hedging flow, such that for the resulting flow $X_t + W_t$ we obtain;

$$x^{(2)} \leq X_t + W_t \leq x^{(1)} \tag{3.1}$$

Then:

- if $X_t > x^{(1)}$, the sign of W_t is opposite to the sign of X_t (and hence W_t is an inflow);
- if $X_t < x^{(2)}$, W_t has the same sign of X_t (and hence W_t is an outflow).

In particular, if $x^{(1)} = x^{(2)}$ from (3.1) we trivially find $X_t + W_t = $ const., so that the random flow stream $X_t, t = 1, 2, \ldots$, is "replaced" by a deterministic flow stream according to the *swap* logic: the owner of the risky outflow stream X_t actually "sells" the stream itself while buying an outflow stream with a sure and constant value.

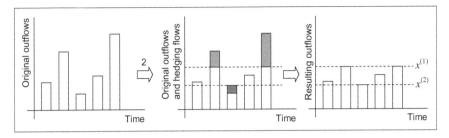

Fig. 3.7 Hedging strategies (2)

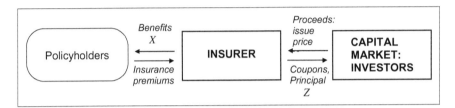

Fig. 3.8 Hedging via mortality bond: a simplified structure

Results of hedging strategies 1(a) and 1(b) are sketched in Fig. 3.6, while results of strategy 2 are shown in Fig. 3.7. Some examples follow.

Strategy 1(a) can be implemented via *insurance*. To reduce the variability of losses, large total losses in some years, X_t, are offset by the benefits Y_t paid by the insurer (an indemnity, or an expense reimbursement). In general, we have $Y_t = \Gamma(X_t)$, where the function Γ expresses the policy conditions, and $Y_t \leq X_t$. In particular, the deductible must be stated consistently with the cap x'. We note, however, that net outflows might be raised because of a possible upper limit in the policy conditions. Anyway, an additional outflow for premium payments must be taken into account.

As regards RM in insurance (and reinsurance), an example of strategy 1(b) is provided, in the framework of Alternative Risk Transfers (ARTs), by *insurance-linked securities* (*ILS*) issued by an insurer (more commonly by a reinsurer) and sold to investors in the capital market. In particular, the impact of catastrophic events (see Sect. 4.5.7) can be hedged by ILS. The payoff, Z_t (in terms of either coupons or principal at maturity, or both), of this type of security is lower when the benefits X_t paid by the insurer (reinsurer) are higher. Mortality bonds, which are described in detail in Sect. 4.5.8, belong to the category of insurance-linked securities. A very simplified structure of this arrangement is depicted in Fig. 3.8.

Again regarding RM in insurance, strategy 1(b) can also be a consequence of a particular business structure. For example, an insurer selling both contracts providing benefit in the case of survival (e.g., life annuities) and contracts paying benefits in the case of death can "automatically" hedge, to some extent, unexpected changes in mortality. Higher amounts X_t paid as life annuity benefits, due to an unanticipated raise in longevity, can be offset by smaller amounts Z_t paid as death benefits, and viceversa. Such a hedging arrangement, as implied by the business structure and not relying on "external" hedging flows, is called *natural hedging*.

A sequence of payments whose amounts are linked to floating interest rates (e.g., instalments amortizing a loan) can suggest the adoption of strategy 2, in order to mitigate the impact of possible changes in interest rates. Upper and lower bounds of the outflow stream depend on interest rate cap and floor (which in turn determine the interest rate *collar*).

Another example of strategy 2 is provided by longevity swaps, which will be described in Sect. 4.6.6.

Of course, whatever the strategy adopted (except for natural hedging), a cost has to be paid by the owner of the risky outflow stream X_t, to transfer (part of) the risk inherent in X_t. It should also be noted that, if strategy 1(b) or 2 is implemented, the owner of the random outflow stream must waive the "acceptable" original outcomes of X_t (that is, $[X_t = \text{low}]$).

3.5 Risk Transfers

Results achieved throughout the risk and impact assessment phases should provide the risk manager with data supporting important decisions, and, in particular:

1. what risks can be retained and what risks should be transferred;
2. how to finance potential losses produced by the retained risks;
3. what kind of risk transfer should be chosen.

It should be stressed that the expression "risk transfer" must obviously be understood as referred to the transfer of possible monetary consequences of the risk (that is, the related impact) and not to a "physical" transfer of the risk itself to another organization.

3.5.1 Retention Versus Transfer

As regards point 1 above, basic guidelines for the decision can follow a *frequency-severity logic* as sketched in Fig. 3.9 in terms of a risk heat map. Risks generating potential losses with low or very low severity (i.e. losses which can be faced thanks to the financial capacity of the firm) can be retained (that is, "self-insured"). In particular, regarding point 2, if the frequency of occurrence is low or very low (for example: risks 1, 2, and 3), the losses do not constitute an important concern and thus can be financed either via internal resources, or via external funds, i.e. borrowing money. Internal resources consist of current cash flows produced by ordinary activities, and shareholders' capital (namely, the assets exceeding the liabilities). A high or very high frequency of losses, on the contrary, suggests funding in advance via specific capital allocation (for example: risks 7, and 8).

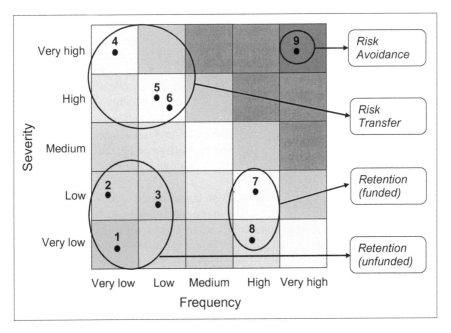

Fig. 3.9 How to manage risks according to their possible impact

Risks generating potential losses with a low or very low frequency but a high or very high severity should be transferred, in particular to an insurance company (for example: risks 4, 5, and 6). Activities implying potential losses with high or very high frequency and high or very high severity should be avoided, because of possible dramatic costs, which may likely lead to bankruptcy (for example: risk 9).

As regards point 3, we note that the term "transfer" should be understood in a rather broad sense: first, it simply denotes "not a full retention" of the risk; secondly, various counterparts, i.e. agents taking (part of) the risk, can be involved to this purpose (see Sect. 3.5.2).

Risks can be partially transferred and, more precisely, only the heaviest part of a potential loss can be transferred whereas amounts which can be faced thanks to the financial capacity of the business can be retained. In particular, the rationale of a risk transfer involving an insurer is the splitting of losses into two parts, one retained by the insured and the other reimbursed by the insurer. More details, related to reinsurance in life insurance business, can be found in Sects. 4.5.3 and 4.5.4.

3.5.2 Counterparts in a Risk Transfer Deal

As regards the counterparts in a risk transfer arrangement, referring to a non-insurance activity, we note what follows. The usual risk transfer involves, as the counterpart, an insurance company (or even more insurance companies). In its turn, an insurance company can transfer risks via appropriate reinsurance arrangements. In the practice of ERM, a deep analysis of all the available insurance (or reinsurance) opportunities should be performed. Convenient insurance covers should be chosen for each type of risk (fire, third-party liability, and so on) borne by the organization. The ultimate result is the construction of an *insurance programme*, possibly involving several insurance companies.

Despite the prominent importance of insurance (and reinsurance) arrangements, other transfer solutions are feasible. For example, large organizations (and, in particular, insurance and reinsurance companies) can transfer risks to capital markets (see below).

A particular form of risk transfer relies on the so-called *captive insurers*. Some large corporations have established their own insurance companies, namely the "captives", to fulfill insurance requirements of various companies inside the group. The captive insurer can be interpreted as a profit center within the group itself. It is worth noting that an insurance company cannot be regarded as a captive simply because completely owned by one or more companies within the group. Conversely, the discriminating feature is whether the majority of its insurance business comes from the companies of the group, rather than from the market or from companies outside the group.

Another solution to risk transfer problems is provided by *pools* which share the same type of risks (without resorting to an insurance company). Examples can be found in professional associations which build up pools to manage specific types of risk, like those related to personal accidents and medical expenses. It should be noted that, as these pools do not imply the existence of an insurance company, their establishment and scope are subject to constraints stated by the relevant legislation.

Finally, we mention the capital market as a possible counterpart in a risk transfer deal. Risk transfers to capital markets, realized by issuing specific bonds, can be placed in the framework of Alternative Risk Transfers (shortly ARTs). This topic, specifically focusing on life annuity and life insurance business, will be addressed in Sects. 4.5.7 and 4.6.4.

3.6   Monitoring: KPIs and the "Time Table"

The "never-ending" feature of the RM process, which clearly appears in Fig. 3.1, implies that the process implementation originates a cycle, which calls for well defined time tables.

Re-identification, re-assessment, etc. of risks are triggered by the findings in the *monitoring* phase. As already noted, this phase has many purposes, in particular:

1. checking whether data and hypotheses assumed in building the model (see Sect. 3.2), e.g., statistical data "imported" from other realties, are close to the organization own experience;

2. checking the effectiveness of the actions undertaken by the organization;
3. determining whether changes in the scenario suggest novel solutions.

Changes in scenario (e.g., related to legislation, fiscal policy, and so on) constitute *exogenous risks*. These changes may occur from time to time, and their occurrence will trigger a new cycle of the RM process.

Actions undertaken by the organization are driven by data and assumptions underlying the construction of the model, and determine the results achieved by the organization.

Results of interest can be summarized by a number of metrics. *Key Performance Indicators* (*KPIs*), which can either refer to the organization as a whole or to specific lines of business, constitute an effective and commonly used metrics.

At the organization level, examples of KPIs are given by the ROE (Return on Equity), and the Net Profit Margin, defined as the ratio of net profits to revenues. The latter is also frequently used referring to a specific line of business.

The multi-year features of life insurance business call for specific KPIs allowing for appropriate time horizons. For example, an indicator of (expected) performance is given by the annual ratio of the Value of New Business (expected present value of profits generated by the policies written in the year) to the total amount of single (or "equivalent") premiums cashed in the year. The above indicator can either refer to a specific line of business or to the insurance company as a whole.

Ratios between actual results and expected results, shortly A/E ratios, can suggest various KPIs yielding an effective monitoring.

Analyzing the KPIs values over time allows us to quantify the performance time-profile. However, diverse results expressed in terms of KPIs, in particular in terms of A/E ratios, may require diverse time frames, depending on the availability and reliability of data. Indeed, to obtain reliable values of the numerator A, some results may call for long time intervals. Hence, diverse time tables may be required.

Interesting examples can be found in the context of life insurance and life annuities. While market results (e.g., the market share) can be monitored rather frequently (on an annual basis, say), mortality experience may require longer periods, in particular when small insurance portfolios are involved. Even longer periods are needed to monitor longevity issues in life annuity portfolios (or pension funds), and to capture possible unanticipated changes in trends: in fact, longevity trends only emerge in multi-year time frames.

Remark It is important to stress the difference between KPIs and KRIs (Key Risk Indicators). KPIs provide an overview of the performance of an organization (and possibly of the various lines of business), by focusing on results already achieved.

Conversely, KRIs (see Sect. 3.2.2) aim at providing information on emerging risk exposures which might affect future results. Although based on past experience, especially in terms of trends, a set of KRIs mainly work as an "early warning system".

3.7 A Forerunner: The Actuarial Control Cycle

We conclude this chapter addressing an approach, proposed and developed in the insurance-actuarial framework, which can be looked at as a forerunner of the ERM logic and, in particular, of the RM process implemented as a cycle: the Actuarial Control Cycle (ACC). Originally proposed in UK by Jeremy Goford in the Eighties, the ACC was then developed in particular by Australian actuaries (see the relevant references in Sect. 3.8).

Focusing on the original contribution by Goford (1985) and referring to a life insurance portfolio, we note what follows (see Fig. 3.10).

According to a set of *initial assumptions* (investment yield, mortality rates, lapse rates, expenses, premium rates, etc.), a *profit test* is performed to calculate the expected cash flows in future years (premiums received, investment income, death and survival benefits paid, surrender values paid, expenses incurred, etc.), and hence to check in particular whether premium amounts are consistent with an assigned

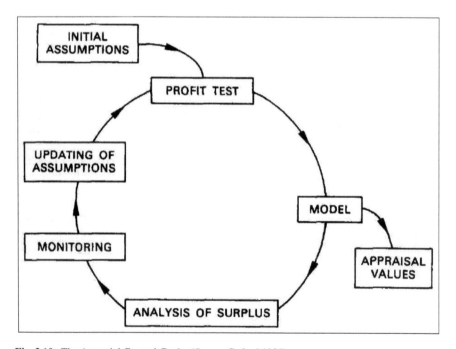

Fig. 3.10 The Actuarial Control Cycle (*Source* Goford 1985)

profit target. The need for a change in the premium rates can be the result of the profit test.

A *model* of the insurance portfolio is then built up and implemented with the data provided by the profit test. Hence, expected cash flows can be generated in relation to both the existing portfolio and the new business as well.

The actual results of the portfolio are then compared to the results provided by the model, and hence the *analysis of surplus* can be performed. The differences between actual and expected results are checked, and the *monitoring* phase can suggest *updating of assumptions* (e.g., investment yield, mortality rates, etc.).

A spin-off result of the control cycle is given by the *appraisal value* calculation: expected present values of future profits can indeed be determined thanks to the model.

Significant analogies with the RM process clearly appear. Nevertheless, it is worth noting that the original ACC model works according to a deterministic setting, as only expected values are addressed. Conversely, appropriate stochastic models leading to the construction of probability distributions of results of interest (e.g., profits) are an essential ingredient of current implementations of the RM process (as we will see in Chaps. 4–7).

3.8 References and Suggestions for Further Reading

Several parts of this chapter are derived from Chap. 1 of Olivieri and Pitacco (2015). The reader can refer to this textbook for a detailed presentation of some issues, like risk pooling and risk transfer via insurance. Further, the reader can refer to most of the books, papers and reports cited in Sect. 2.3.

Own Risk and Solvency Assessment (ORSA), as the core of ERM, is addressed in the IAA Risk Book by Coggins et al. (2016), highlighting the role of the actuary in creating valuable risk analysis frameworks.

While models are needed in risk assessment and impact assessment phases of the RM process, the user is exposed to the model risk. Appropriate model governance is then required. These topics are specifically focused by Howes et al. (2019) in the IAA Risk Book.

A vast literature addresses Key Performance Indicators (KPIs). We only cite a few contributions. The book by Parmenter (2015) provides an extensive presentation of advantages and disadvantages in the use of KPIs. Guidelines for using KPIs can be found in PwC (2007), while performance in life insurance companies is focused by Kirova and Steinmann (2012).

Key Risk Indicators (KRIs) constitute a metrics for expressing risk impacts and a communication tool inside the organization. Distinguishing KPIs from KRIs is the starting point of Beasley et al. (2010) which then addresses the development of effective KRIs. Branson (2015) focuses on communication of key risk information. Interesting case studies are presented by Boyd et al. (2016) and Chou et al. (2020).

The Actuarial Control Cycle was originally proposed by Goford (1985). Various ideas underpinning the original proposal have then be developed and implemented. For a detailed analysis, the reader is referred, for example, to Bellis et al. (2010) and Gribble (2003).

Chapter 4
RM for Life Insurance and Life Annuities

4.1 Risk Identification

In the risk identification phase of the RM process, we have to single out *risk causes* and *risk components* and, looking at impact on results of interest, *risk factors*.

4.1.1 Risk Causes and Risk Factors: An Introduction

Risk causes and risk components in life insurance will be analyzed in detail in Sect. 4.2, while risk factors will be addressed in Sect. 4.4.1. In this section, we only provide some examples.

Random quantities, which constitute risk causes in a life insurance portfolio, are for example the following ones:

- number of deaths (mortality of insureds);
- yield from investments (backing mathematical reserves and shareholders' capital);
- future expenses;
- inflation;
- number of lapses, policyholders' options, etc.

Some random quantities are correlated. For example: inflation, investment yield, number of surrenders.

Risk factors can be singled out by looking at characteristics of an insurance portfolio, whose (quantitative or qualitative) values determine a higher or lower impact for each risk cause.

For example, a portfolio mainly consisting of term insurance policies is particularly exposed to the risk cause given by insureds' mortality. Conversely, a portfolio with a significant number of endowment insurance policies is more exposed to financial risks.

© Springer Nature Switzerland AG 2020
E. Pitacco, *ERM and QRM in Life Insurance*, Springer Actuarial,
https://doi.org/10.1007/978-3-030-49852-8_4

4.1.2 Risk Components

Refer to a given result from a business, e.g., the profit as at the end of a given year from a life insurance portfolio. The result is, of course, random and can, in particular, be summarized by its expected value. Then, compare the actual result to the expected one. Why may the actual result differ from its expected value? The analysis of risk components can provide the answer.

(a) Ordinary *random fluctuations* of a quantity around its expected value constitute the risk component usually called *volatility*, or *idiosyncratic risk*, or *process risk*. This risk component arises at an individual level, being caused, for example, by individual random lifetimes.

(b) *Systematic deviations* from the expected value constitute the second risk component. This risk component is frequently named *uncertainty risk*, because originated by uncertainty in the representation of a random phenomenon. The representation might not properly reflect the features of the phenomenon; for example, the assumed life table might not represent the specific mortality of the insureds. If the representation is based on a model and the relevant parameters, then we may have:

- *model* risk; for example, the mortality is assumed to follow the Gompertz law, whilst the age-pattern of mortality is not exponential;
- *parameter* risk; the chosen model correctly represents the shape of the age-pattern of mortality, but a bad estimation affects the parameters.

Systematic deviations affect the pool of individuals as an aggregate, as the uncertainty in representing the random phenomenon concerns all the individuals in the pool.

(c) Extreme events (or "tail events"), i.e. events with a very low probability but a tremendous financial impact, constitute the *catastrophe risk* component. For example, a sudden diffusion of a contagious disease may cause and exceptional raise in mortality.

Figures 4.1, 4.2, 4.3, 4.4, 4.5, 4.6 and 4.7 illustrate the effect of the three risk components on the frequency of death (represented by dots) at a given age x, compared to the one-year probability of death at that age (represented by the dashed line). In particular, Figs. 4.1, 4.2 and 4.3 refer to probability q_x assumed constant throughout time (an example of "level case"), while Figs. 4.4, 4.5, 4.6 and 4.7 refer to probability $q_x(t)$ assumed decreasing throughout time (an example of "trend case").

As regards the risk of systematic deviations, it is interesting to compare the situations plotted in Figs. 4.5 and 4.7, respectively. The former represents a case of overestimation of the mortality improvements, although the "slope" of the assumed trend of the probability of death $q_x(t)$ corresponds to the slope of the actual trend (the dotted line). The latter, on the contrary, represents a case of overestimation of the decreasing trend in the probability $q_x(t)$: the assumed decrease is indeed stronger than the actual one.

Fig. 4.1 Level case (a)

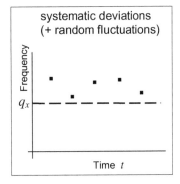

Fig. 4.2 Level case (b)

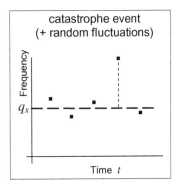

Fig. 4.3 Level case (c)

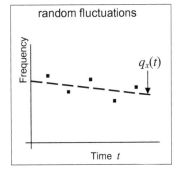

Fig. 4.4 Trend case (a)

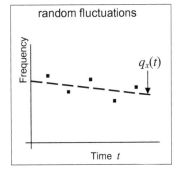

Fig. 4.5 Trend case (b)

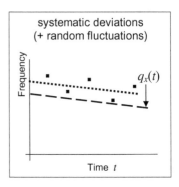

Fig. 4.6 Trend case (c)

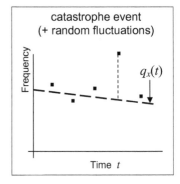

Fig. 4.7 Trend case (b')

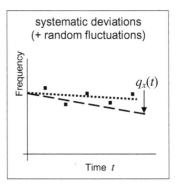

Remark 4.1 The terms *risk* and *uncertainty* have been used with diverse meanings. In particular, "risk" is frequently used when referring to phenomena for which some probabilistic model can be constructed and estimated, also relying on statistical evidence (e.g., the risk of ordinary random fluctuations). Conversely, "uncertainty" usually refers to difficulties in representing a phenomenon via an appropriate probabilistic model. In the present lecture notes, the expression "uncertainty risk" denotes the latter situation.

Remark 4.2 The risk of systematic deviations, sometimes in the shortened version "systematic risk", should not be confused with the *systemic risk*. The expression systemic risk is used to denote the collapse of an entire financial system or an entire market. The systemic risk is a consequence of interdependencies in a system or market, where the failure of a single company, institution, etc. can cause, via propagation, the failure of many other organizations, thus triggering a system or market implosion.

4.1.3 More on Risk Components: Hedging the Impact

The three risk components defined in Sect. 4.1.2 imply diverse impacts and hence call for specific RM actions. Although impact assessment and hedging actions constitute topics of following sections, some ideas can conveniently be presented here to stress the diversity of the three risk components.

We refer, for simplicity, to the annual payout of a life insurance portfolio consisting of term insurance policies which will pay death benefits. For simple formal proofs of the results below, the reader is referred to Olivieri and Pitacco (2015).

We assume that the impact of the mortality risk is expressed, in "relative" terms, by the *risk index* (or *coefficient of variation*) ρ, defined as the the standard deviation of the portfolio payout over the expected value of the payout itself.

(a) If the portfolio payout is only affected by mortality random fluctuations, it can be proved that, under rather general conditions, the risk index decreases as the portfolio size increases. In particular, for a perfectly homogeneous portfolio (both in terms of sum assured and probability of death), the risk index can be expressed as follows:

$$\rho = \sqrt{\frac{A}{n}} \tag{4.1}$$

where n is the portfolio size, and A only depends on the probability of death. Hence, this risk component can be faced by increasing the portfolio size, or, if an increase is impossible because of a limited market share, by reinsuring the portfolio payout. The reinsurer, thanks to a larger portfolio size, can better face this risk component. The strategy based on increasing the portfolio size is usually referred to as *diversification via pooling*.

(b) Systematic deviations cannot be faced via pooling. Indeed, as previously noted, this risk component affects the pool as an aggregate. Its total impact on portfolio results increases as the portfolio size increases. Although increasing the portfolio size implies a reduction in the risk index, a term of the index itself does not depend on the portfolio size and hence represents the non-diversifiable part of the risk of systematic deviations. Formally, for a homogeneous portfolio, instead of the result expressed by Eq. (4.1) we find:

$$\rho = \sqrt{\frac{B}{n} + C} \tag{4.2}$$

where C is independent of n. Different strategies are then required, such as hedging via Alternative Risk Transfers (ARTs; see Sects. 4.5.6 and 4.5.7).
(c) The catastrophe risk component can first be managed via an appropriate geographic diversification, thus avoiding significant concentration of insured risks in geographic areas particularly exposed to catastrophes (e.g., floods, hurricanes, earthquakes, etc.). Reinsurance and ARTs provide further hedging.

Remark As noted by Haberman (1996), pioneering contributions to the analysis of effects of the pool size on the risk profile were provided in the 18th century by Corbyn Morris and Johannes N. Tetens. In particular, Tetens in his book dated 1786 noted that, while the variance of the pool total payout increases with the pool size, the risk (expressed by the standard deviation) in respect of each individual member of the pool reduces when the pool size becomes larger, the square root of the size n playing a substantial role (see Eq. (4.1)). It is worth stressing that the above contributions can be considered milestones in the analysis of the portfolio risk profile.

4.2   Risk Causes and Risk Components in a Life Insurance Portfolio

A number of classifications have been proposed by national and international institutions, for example the US NAIC (National Association of Insurance Commissioners), and the IAA (International Actuarial Association). A classification has also been proposed in the framework of Solvency II. In this section we will follow the classification proposed by the Solvency Working Party of the IAA, in order to define solvency standards (see IAA (2004)).

The following categories of risk causes have been identified:

1. Underwriting risks;
2. Credit risks;
3. Market risks;
4. Operational risks;
5. Liquidity risks;
6. Event risks.

As we will see in the following sections, overlaps can be detected. Moreover, correlations among various risk causes should be considered.

4.2.1   Underwriting Risks

This category includes all the risks taken by an insurance company as a consequence of its underwriting activity.

The following causes of risk are included.

- All risks originated by the insureds' random lifetimes are denoted as *mortality/longevity risks*. The three risk components (see Sect. 4.1.2) can be recognized:

    - random fluctuations around the expected age-pattern of mortality;
    - systematic deviations from the expected age-pattern of mortality;
    - catastrophe risk.

Remark 4.1 According to the current insurance language:

- *mortality risk* denotes the risk of mortality higher than expected; it is a downside risk when insurance products with positive sum at risk (given by the benefit in case of death less the policy reserve) are involved, e.g., term insurance and endowment insurance;
- *longevity risk* denotes the risk of mortality lower than expected; it constitutes a downside risk for insurance products with negative sum at risk, e.g., life annuities, pure endowments; in particular, the expression *aggregate longevity risk* is used to refer to the systematic deviations component.

- *Disability risks* are originated by the insureds' health status; the expression is commonly used to denote both working disability as well as senescent disability (the latter interesting in particular long-term care insurance products).

Remark 4.2 The expression *biometric risks* usually encompasses mortality, longevity and disability risks.

- The expression *policyholders' behavior risk* (or *option risk*) encompasses all the risks caused by policyholders' possible choices: surrender, lapse, exercise of the annuitization option, extension of the policy duration (with a guaranteed interest rate), etc. Policyholders' choices may have direct and indirect impacts. For example:

    - lapses have, as a direct consequence, difficulty in recovering acquisition expenses;
    - a raise in longevity risk is the indirect consequence of the exercise of the annuitization option.

- *Pricing risk* denotes the possibility of losses because of poor premium levels, for example due to:

    - non appropriate technical bases;
    - non appropriate pricing of guarantees and options embedded in the insurance product;
    - non appropriate expense loading parameters.

- *Selection risk* is the risk of losses because of an uncorrect assessment of insureds' characteristics, for example in case of underestimation of the extra-mortality of substandard lives. See also Sect. 4.2.7.

4.2.2 Credit Risks

Generally speaking, the expression *credit risks* denotes a rather broad category of risks. For all the investors who purchased bonds, credit risk denotes the risk of insolvency of bond issuers.

For an insurance company, besides the risk of insolvency of bond issuers this category also includes the risk associated with the failure of reinsurers to meet their obligations when due.

4.2.3 Market Risks

Market risks (also denoted as *performance risks*, or *financial risks*) are all the risks introduced into insurance company activity through variations in financial markets, measured by changes in interest rates, equity indices, etc.

Remark In some classifications, the expression *investment risks* is used to encompass both market risks and credit risks.

In particular, risk of losses may be due to variations in interest rates (*interest rate risk*), in equity prices and stock exchange indices (*equity risk*), in currency exchange rates (*exchange rate risk*).

In this context, the expression *basis risk* denotes the risk of variations in asset value different from variations in liability value. For example, this risk arises because of assets backing the reserves of unit-linked policies different from the assets which constitute the reference fund used to determine the unit value.

The *ALM (Asset-Liability Management) risk* is caused by variations in interest rates and equity prices which might have different impacts on insurer's assets and liabilities.

Market risks include all the three components (see Sect. 4.1.2):

- process risk (or volatility), that is, the natural variability in interest rates and equity values;
- uncertainty risk, originated by difficulty in finding an appropriate financial model (e.g., does the Brownian motion provide an appropriate representation of random yields?), and in the relevant parameter estimation;
- catastrophe risk, because of tail events, for example the economic collapse of 1929.

4.2.4 Operational Risks

The concept of *operational risks* comes from the banking industry, where it has been adopted to denote all the risks other than credit risk and market risk. Generally speaking, these risks are related to:

- human resources;
- processes, in particular:

  - accounting;
  - administration;
  - planning;

- data processing.

4.2.5 Liquidity Risks

Liquid assets are needed to meet the cash flow requirements. The term *liquidity risks* denotes the risk of losses in the event of insufficient liquid assets. In particular, the following risks are included in this category:

- *asset liquidation risk*, that is, the risk that unexpected timing or amounts of cash needed may require liquidation of assets when market conditions might result in loss of realized value;
- *capital market risk*, that is, the risk that the insurance company will not be able to obtain sufficient funding sources from outside markets.

The presence of liquidity risks calls for an appropriate *liquidity management*, which must be performed at various levels, in particular:

- day-to-day cash management, i.e. a treasury function;
- on-going cash flow management, that is, monitoring cash needs for the next 6–24 months.

Further, special attention should be placed on the catastrophe component of the liquidity risks: a huge liquidity need might typically occur because of an extraordinary number of surrenders, e.g., motivated by a sudden change in the market conditions. Such a risk should preliminarily be avoided (or, at least, mitigated) via an appropriate product design and, in particular, via policy conditions.

4.2.6 Event Risks

This category encompasses all the risks which are outside the control of the insurance company. Examples are as follows:

- *reputation risk*;
- *legal risks*, usually arising from contracts which are not enforceable or documented correctly;
- *disaster risk*, beyond possible direct consequences on insurer's obligations;
- *regulatory risk*, due to sudden changes in the regulatory environment;
- *political risks*, due to changes in the political and economical scenario.

4.2.7 Selection Risk: Further Issues

Selection risk has been defined in Sect. 4.2.1. As noted, losses might result in a portfolio of policies providing death benefits because of underestimation of insureds' mortality, in particular because of underestimation of the extra-mortality of substandard insureds. Extending our scope to health insurance products (for example, sickness insurance covers, insurance products providing annuities in case of disablement, etc.), we note that losses might occur in case of wrong ascertainment of insured's (current and past) pathologies and exposure to injuries, and related consequences in the use of the health insurance policy.

Selection risk arises because of *information asymmetry*. This expression generally denotes a market situation in which buyers and sellers have different information about features of a transaction. In insurance (in particular, in life and health insurance), the individual risk profile consists of:

- *observable risk factors* (e.g., the individual's current health status);
- *unobservable risk factors* (e.g., future behavior of the insured regarding the use of the insurance cover).

Of course, underwriting can only rely on observable risk factors, and this causes information asymmetry. Accurate underwriting improves insurer's information and then reduces information asymmetry. Nonetheless, residual information asymmetry implies *adverse selection* risk and *moral hazard* risk.

The expression "adverse selection" encompasses all the situations in which an individual's demand for insurance is positively correlated with his/her risk exposure (beyond the level which can be ascertained via underwriting). For example:

- individual mortality higher than that assumed or assessed via underwriting;
- individual need for care higher than that assumed or assessed via underwriting.

Adverse selection risk can be increased by the *anti-selection* effect, that is the individual propensity to buy insurance because of a perceived risk exposure (e.g., health risk exposure) higher than that accounted for by the insurer in calculating the premium.

Information asymmetry regarding individual health status originates adverse selection. Conversely, the expression "moral hazard" refers to a situation regarding the insured's behavior, that is, the use of the insurance cover. For example, moral hazard in health insurance occurs when:

- an insured accepts more risky health situations, and then use more health care because aware of the cost transfer to the insurer;
- an insured uses the insurance cover even for light disease situations usually not requiring specific care and hence insurer's intervention;
- a (previously) disabled insured prolongs cashing the disability annuity benefit beyond his/her recovery.

4.3 Risk Assessment

In the risk assessment phase of the RM process risk causes and related components are expressed in quantitative terms via appropriate stochastic models.

4.3.1 Assessing Risks in a Life Insurance Portfolio

Life insurance and life annuity portfolios are affected by numerous risk causes (and related components). A map of risk causes has been presented in Sect. 4.2. Different risk causes call for different modeling structures. Consider the following examples.

- Mortality and longevity risks must first be quantified by assumptions on the age-pattern of mortality, in terms of a life table or a mortality law. Random fluctuations of the numbers of deaths around the relevant expected values can be modeled after assuming hypotheses about independence or correlation among individual lifetimes. Further, uncertainty about mortality assumptions can be expressed, for example, by defining a set of mortality scenarios; so, the risk of systematic deviations can be quantified.
- Assessing market risks calls for models representing, in particular, random yields from investments backing the insurer's liability and the shareholders' capital.

Several examples will be presented in Chaps. 6–8. In the following sections we only provide some basic issues on mortality modeling.

4.3.2 Deterministic Models Versus Stochastic Models[1]

Assume that, at time $t = 0$, a "group" (for example a pension fund, or a portfolio of life insurance contracts) initially consists of a given number, N_0, of individuals, all aged x_0. Further, assume that no other individual will later enter the group. Thus, the group is a cohort. Finally, assume death as the only cause of exit.

The number of people alive at time t, $t = 1, 2, \ldots$, is a random number, which we denote by N_t. Any sequence of integers $n_1, n_2, \ldots$, such that

$$(N_0 \geq) \, n_1 \geq n_2 \geq \ldots \tag{4.3}$$

is a possible outcome of the random sequence

$$N_1, \ N_2, \ \ldots \tag{4.4}$$

---

[1] Sections 4.3.2–4.3.4 are based on the contents of Sect. 3.10 of Olivieri and Pitacco (2015).

Of course, any single outcome of the random sequence (4.4) does not provide, by itself, significant information about the reasonable evolution of the cohort. Conversely, the meaning of "reasonable" can be specified as soon as a probabilistic structure describing the lifetimes of the cohort members has been assigned.

We assume that the individuals in the cohort are analogous in respect of the age-pattern of mortality, thus for all the individuals we assume the same life table (or mortality law). Hence, for any member of the cohort, the probability of being alive at time t is given (according to the usual actuarial notation) by $_t p_{x_0}$.

It follows that the expected number of individuals alive at time t, out of the initial N_0 members, is given by:

$$\mathbb{E}[N_t] = N_0 \, _t p_{x_0}; \quad t = 1, 2, \ldots \tag{4.5}$$

It should be noted that, although formula (4.5) involves probabilities, the model built up so far is a deterministic model, as probabilities are only used to determine expected values and the probabilities themselves are assumed to be known. A first step towards stochastic models follows.

We assume that the random lifetimes of the individuals in the cohort are independent. For any given t and for $j = 1, 2, \ldots, N_0$, we denote by $\mathscr{E}_t^{(j)}$ the event "the member j is alive at time t". Of course, $\mathbb{P}[\mathscr{E}_t^{(j)}] = {}_t p_{x_0}$ for all j. From the independence of the lifetimes, the independence of the events $\mathscr{E}_t^{(j)}$, $j = 1, 2, \ldots, N_0$, follows. We note that N_t can be defined as the random number of true events out of the N_0 events defined above; hence, N_t has a binomial distribution, with parameters $N_0, \, _t p_{x_0}$. Thus,

$$\mathbb{P}[N_t = k] = \binom{N_0}{k} (_t p_{x_0})^k (1 - {}_t p_{x_0})^{N_0 - k}; \quad k = 0, 1, \ldots, N_0 \tag{4.6}$$

In particular, the variance of N_t is given by:

$$\mathbb{V}\mathrm{ar}[N_t] = N_0 \, _t p_{x_0} (1 - {}_t p_{x_0}) \tag{4.7}$$

and the risk index (or coefficient of variation) by:

$$\rho[N_t] = \frac{\sqrt{\mathbb{V}\mathrm{ar}[N_t]}}{\mathbb{E}[N_t]} = \sqrt{\frac{1 - {}_t p_{x_0}}{N_0 \, _t p_{x_0}}} \tag{4.8}$$

Example 4.1 We consider a cohort of N_0 individuals, all age $x_0 = 40$ initially. We assume that the age-pattern of mortality is described by the first Heligman-Pollard law,[2]

$$\frac{q_x}{1 - q_x} = A^{(x+B)^C} + D \, e^{-E(\ln x - \ln F)^2} + G \, H^x \tag{4.9}$$

---

[2] See: Heligman and Pollard (1980).

Table 4.1 Parameters of the first Heligman-Pollard law

A	B	C	D	E	F	G	H
0.00054	0.01700	0.10100	0.00016	10.72	18.67	1.83 E−05	1.1100

Table 4.2 Some markers of the first Heligman-Pollard law

$\overset{\circ}{e}_0$	$\overset{\circ}{e}_{40}$	$\overset{\circ}{e}_{65}$	Mode	q_0	q_{40}	q_{80}
77.282	38.601	16.725	83	0.00684	0.00121	0.07178

Table 4.3 From the probability distribution of N_t ($N_0 = 500$)

t	$\mathbb{E}[N_t]$	$\mathbb{V}\text{ar}[N_t]$	$\rho[N_t]$
5	496.269	3.703	0.003878
10	490.083	9.720	0.006362

Table 4.4 From the probability distribution of N_t ($N_0 = 1\,000$)

t	$\mathbb{E}[N_t]$	$\mathbb{V}\text{ar}[N_t]$	$\rho[N_t]$
5	992.538	7.406	0.002741
10	980.166	19.441	0.004498

where q_x denotes the one-year probability of death for an individual age x; we assume the parameter values specified in Table 4.1. The marker values shown in Table 4.2 (where "Mode" denotes the modal age at death) follow.

Refer to a cohort initially consisting of $N_0 = 500$ individuals; the probability distributions of N_5 and N_{10} are plotted in Figs. 4.8 and 4.9, respectively. For example, we find the values displayed in Table 4.3.

For a cohort of $N_0 = 1\,000$, the probability distributions of N_5 and N_{10} are plotted in Figs. 4.10 and 4.11, respectively, and we find the values shown in Table 4.4.

Looking at Figs. 4.8 to 4.11 (of course, taking into account the different scales adopted for the axes) and at Tables 4.3 and 4.4, we note what follows.

- For any given initial size N_0, the variability (in both absolute and relative terms) increases over time as the portfolio sizes decreases.
- A larger initial portfolio size N_0 implies a smaller relative variability, thus witnessing the pooling effect.

∎

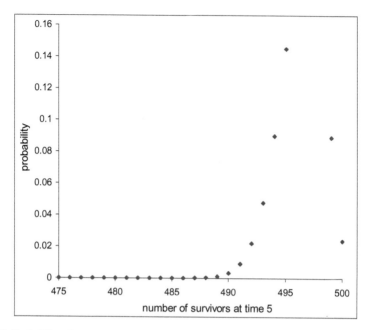

Fig. 4.8 Probability distribution of N_5 ($N_0 = 500$)

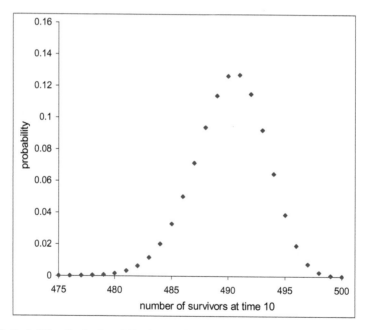

Fig. 4.9 Probability distribution of N_{10} ($N_0 = 500$)

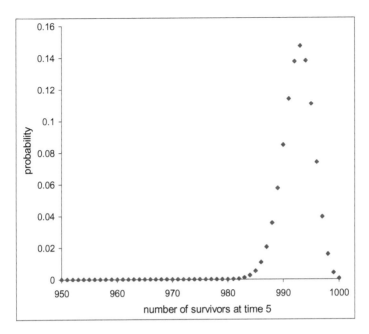

Fig. 4.10 Probability distribution of N_5 ($N_0 = 1\,000$)

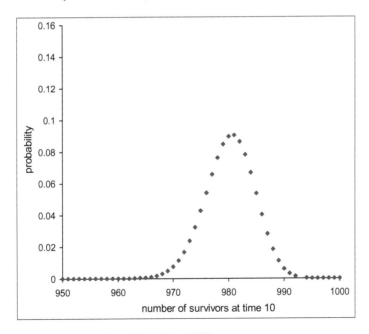

Fig. 4.11 Probability distribution of N_{10} ($N_0 = 1\,000$)

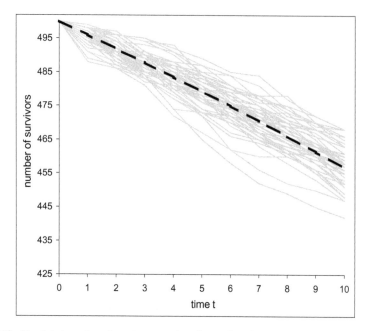

Fig. 4.12 Simulated number of survivors: random fluctuations ($N_0 = 500$)

4.3.3   Random Fluctuations in Mortality

The probability distribution of N_t witnesses the presence of random fluctuations in the number of survivors around its expected value $\mathbb{E}[N_t]$, which are the consequence of the process risk. Further insights into the process risk can be obtained looking at the random behavior of the number of survivors in the cohort over time. As life insurance is, typically, a medium-long term business, the features of this activity can be better perceived from a dynamic perspective.

To this purpose, we can implement a stochastic simulation procedure, based on the generation of (pseudo-) random numbers. The procedure can be as follows:

1. simulate the random lifetime (i.e. the age at death) for each member of the cohort;
2. given the simulated values of the N_0 lifetimes, calculate the numbers of individuals alive at times $1, 2, \ldots$, namely the simulated outcomes $n_1, n_2, \ldots$ of the random numbers $N_1, N_2, \ldots$;
3. repeat steps 1 and 2, for a given number n_{sim} of times.

The output of this procedure is a (simulated) sample consisting of n_{sim} outcomes, or paths, of the random sequence $N_1, N_2, \ldots$.

Example 4.2 We consider a cohort of $N_0 = 500$ individuals, all age $x_0 = 40$ initially. As in Example 4.1, we assume the age-pattern of mortality described by the first Heligman-Pollard law. Figure 4.12 illustrates the behavior of $n_{sim} = 50$ paths of the random sequence

$$N_1, N_2, \ldots, N_{10}$$

namely limited to the first 10 years. The dashed line represents the sequence of expected values

$$\mathbb{E}[N_1], \mathbb{E}[N_2], \ldots, \mathbb{E}[N_{10}]$$

around which the simulated paths develop. ∎

For any given time t, information about the distribution of the random number N_t can be obtained looking at the simulated outcomes of N_t, namely by constructing the statistical distribution of N_t.

However, it is worth noting that, when just one cohort consisting of individuals with the same age-pattern of mortality is involved, probability distributions of the random numbers of survivors can be found via analytical formulae, as seen in Sect. 4.3.2. Further, approximations to the probability distribution of the numbers of people dying in the various years can be adopted, when several initial ages and hence several age-patterns of mortality are involved.

Conversely, simulation procedures are useful, even when the structure by age of the cohort is very simple, when we have to analyze the impact of the mortality risk, that is, the behavior of quantities depending on the random numbers of people alive or dying. Important examples are given by the cash-flows in life insurance portfolios. So, the simulation procedure we have described should be meant as the starting point for building up more complex models involving, for example, incomes and outflows. Examples will be provided in Chaps. 6 and 7.

4.3.4 Systematic Deviations in Mortality

In order to represent the age-pattern of mortality in a given group (namely, a life insurance portfolio or a pension plan), we have to choose a life table or a mortality law. However, the mortality actually experienced by the group in future years may "systematically" differ from the one we have assumed. This may occur for various reasons. For example:

- because of poor past experience, we have chosen a life table relying on mortality experienced in other populations;
- the future trend in mortality differs from the forecasted one (expressed by a projected life table).

So, whatever hypothesis has been assumed, the future level and trend in mortality are random. Then, uncertainty risk arises, namely the risk due to uncertainty in representing the mortality scenario. Hence, systematic deviations from the expected values may occur, which combine with ordinary random fluctuations.

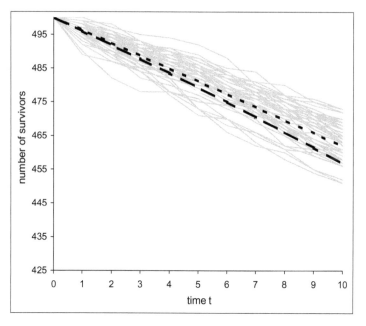

Fig. 4.13 Simulated number of survivors: systematic deviations (and random fluctuations)

Example 4.3 We refer to the cohort already considered in Example 4.2. First, we assume the age-pattern of mortality described by the Heligman-Pollard law with the parameters adopted in Example 4.2. Then, we suppose that the future "actual" mortality follows the Heligman-Pollard law in which the previous values of the parameters G and H are replaced by:

$$\bar{G} = 2.2875\,\mathrm{E}{-05}; \quad \bar{H} = 1.0878$$

We denote by $\mathbb{E}[N_t|G, H]$ and $\mathbb{E}[N_t|\bar{G}, \bar{H}], t = 1, 2, \ldots$, the expected values based on the first and the second assumption respectively. Figure 4.13 illustrates the behavior of 50 paths of the random sequence:

$$N_1, N_2, \ldots, N_{10}$$

(i.e. limited to the first 10 years), simulated according to the new assumption about the mortality. The dashed line represents the expected values

$$\mathbb{E}[N_1|G, H], \mathbb{E}[N_2|G, H], \ldots, \mathbb{E}[N_{10}|G, H]$$

whereas the dotted line represents the expected values

$$\mathbb{E}[N_1|\bar{G}, \bar{H}], \mathbb{E}[N_2|\bar{G}, \bar{H}], \ldots, \mathbb{E}[N_{10}|\bar{G}, \bar{H}]$$

around which the simulated paths develop. The process risk causes the random fluctuations around the $\mathbb{E}[N_t | \bar{G}, \bar{H}]$'s, while the uncertainty risk originates the systematic deviations from the $\mathbb{E}[N_t | G, H]$'s. We note that the expected numbers of survivors according to the first assumption represent an underestimation with respect to the "actual" numbers of survivors.   ∎

4.4 Moving to Impact Assessment

As already noted, complex problems arise when moving from risk assessment to impact assessment, a critical step in the framework of QRM. In this section, we analyze feasible approaches to these problems.

4.4.1 Risk Factors in a Life Insurance Portfolio

As noted in Sect. 4.1.1, some characteristics of an insurance portfolio can determine a higher or lower impact for each risk cause.

In a life insurance portfolio, the prevailing product type raises or lowers the impact of various risk causes. For example, term insurance contracts raise the impact of the mortality risk, while reduce the impact of market risks. Conversely, the impact of market risks is raised by a prevailing presence of endowment insurance policies.

A larger portfolio size improves the diversification effect via risk pooling, regarding the random fluctuation component. Moreover, the lower the dispersion in the distribution of the sums assured, the better the diversification, the optimal portfolio consisting in policies with the same sum assured. Conversely, the larger is the portfolio the more severe is the financial impact of systematic deviations.

Policy conditions can significantly change the impact of some risk causes because of policyholders' behavior. For example, generous surrender conditions can enhance the propensity to surrender, then rising the possible impact of the liquidity risk. The presence of an annuitization option can increase the impact of longevity risk.

The link between risk causes and components on the one hand, and impacts on the other is sketched in Fig. 4.14.

4.4.2 From Scenario to Results

A number of results can be chosen to assess the impact of risks. The choice should be driven by the main purpose of the analysis we are going to perform. In particular:

- *liquidity* analysis calls for focus on cash flows;
- evaluating *profitability* of course requires the assessment of profits;

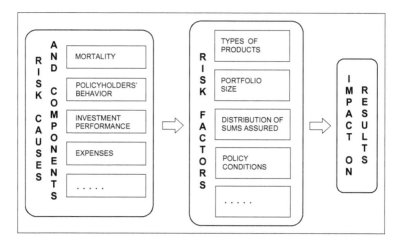

Fig. 4.14 The "role" of risk factors: from risks to impact

- *solvency* analysis is usually based on the amount of the Net Asset Value (NAV).

A detailed analysis over time calls for periodic (e.g., quarterly, or annual) results, whereas simplified assessments can be based on a single result, e.g., at the end of a given period.

The time horizon, through which assessments are performed, may range from short-term (e.g., six months, or a year) to long-term (e.g., thirty years). Asymptotic results can also be evaluated, by resorting to analytical models which, however, require significant simplifications. For this reason, we disregard this type of results.

Finally, a choice concerns how to define the portfolio:

- just in-force policies are taken into account;
- also expected future entrants, that is new business, are included in the assessment.

Types of results which can be assessed are sketched in Fig. 4.15, where risk causes affecting the scenario are grouped according to the classification presented in Sect. 4.2.

4.4.3 A Formal Setting

Whenever we aim at representing the impacts of several risk causes, an appropriate stochastic model is required. Generally speaking, we need to "transform" *input variables* into *output variables*. The transformation can formally be represented as follows:

$$(Y_1, Y_2, \dots) = \Phi(X_1, X_2, X_3, \dots; d_1, d_2, \dots) \tag{4.10}$$

where:

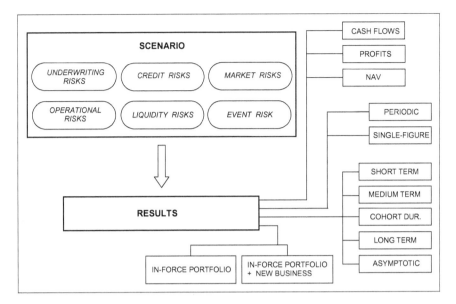

Fig. 4.15 From scenario to results

- $X_1, X_2, X_3, \dots$ denote input random variables, or *scenario variables*, each one related to a risk cause; e.g., with reference to a given year: number of deaths in the portfolio, investment yield, amount of expenses, etc.;
- $d_1, d_2, \dots$ denote input variables under control, or *decision variables*; e.g., the interest rate adopted in premium calculation, a set of parameters to express expense loading, etc.;
- $Y_1, Y_2, \dots$ denote output random variables representing the impact on results of interest, e.g., cash flows, profits, net asset value (NAV) as at the end of the year, etc.;
- Φ is a vector-valued function, frequently called *aggregation function*, which transforms the input variables into output variables.

The logical structure of the above transformation is sketched in Fig. 4.16.

A "complete" stochastic model should in principle transform, for given values of the decision variables, the joint probability distribution of the scenario variables into the joint probability distribution of the results of interest. In practice, this transformation is untractable in terms of closed-form expressions, if a realistic representation of the scenario is provided. Hence some modeling choices are required, as well as some (simplified) implementation techniques must in practice be adopted.

In the following sections we proceed from simple deterministic approaches to a complex double-stochastic setting. We focus on three scenario variables, X_1, X_2, X_3, and one result, Y.

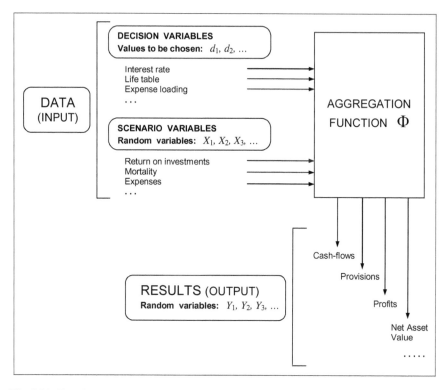

Fig. 4.16 From input to output via the aggregation function

4.4.4   From Deterministic to Stochastic Approaches

Simple deterministic approaches are sketched in Fig. 4.17. In particular, 1a represents the simple calculation of the value of Y corresponding to three chosen values of X_1, X_2, X_3, which can be, for example, expected values or modal values or values of specific interest, e.g., "tail" values (the value of X_3 in Fig. 4.17, 1a) in order to perform *stress testing*. Of course, no information can be obtained about the probability distribution of Y. Nonetheless this approach reveals its importance in checking the consequences of extreme scenarios, for example the impact of a huge number of surrenders on the insurer's liquidity.

Conversely, 1b represents the tabulation of Y against X_1 (and possibly the fitting of the tabulation results). Also this approach does not provide any information about the probability distribution of Y, while can be useful to test the *sensitivity* of results when one or more input variables change value.

"True" stochastic approaches are sketched in Fig. 4.18. The aim is to find the probability distribution of the random result Y, from which several synthetic information can be derived (expected value, variance, value at risk, etc.). In particular, 2a represents the calculation of the conditional distribution of Y as a function of the

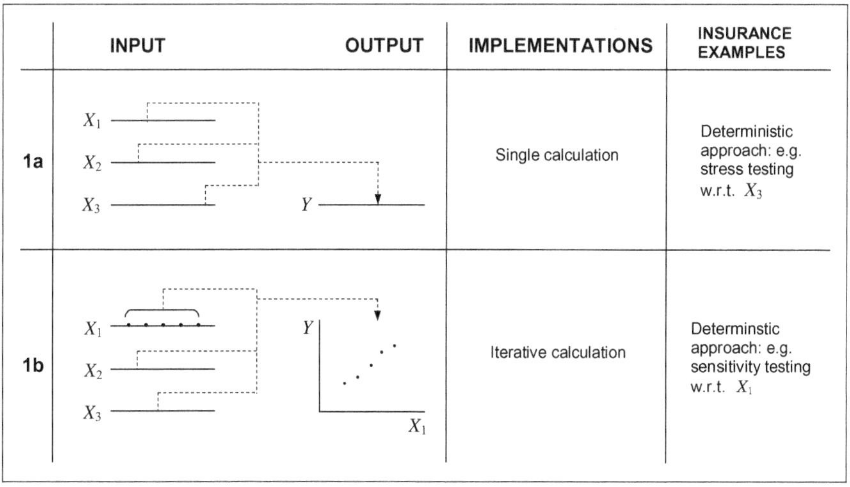

Fig. 4.17   From risk assessment to impact assessment: deterministic approaches

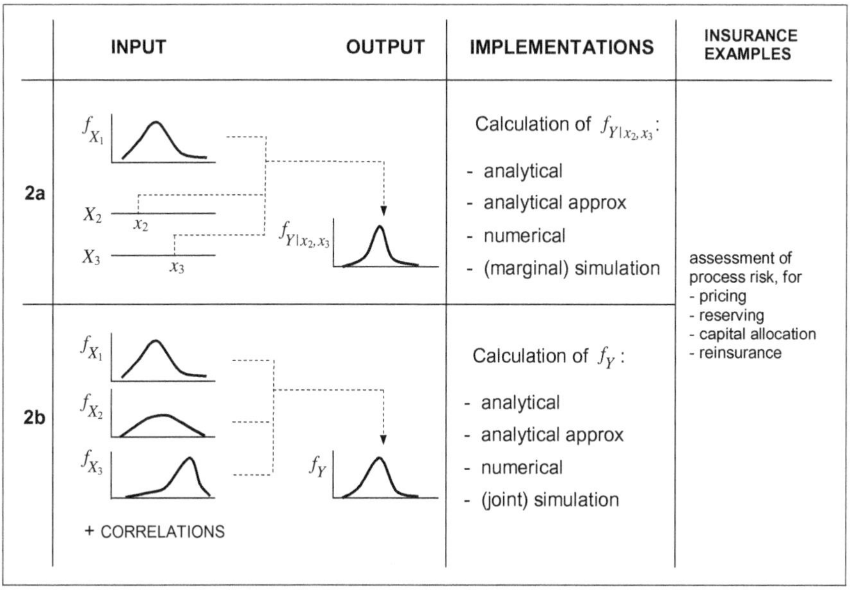

Fig. 4.18   From risk assessment to impact assessment: stochastic approaches

random variable X_1, for given values of the other input variables, that is, $X_2 = x_2$ and $X_3 = x_3$. Conversely, 2b represents the calculation of the unconditional distribution of Y.

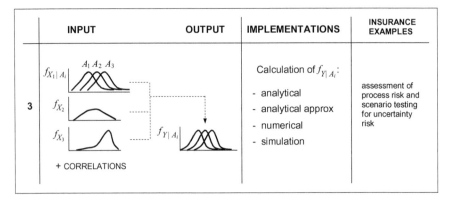

Fig. 4.19 From risk assessment to impact assessment: stochastic approach allowing for uncertainty risk

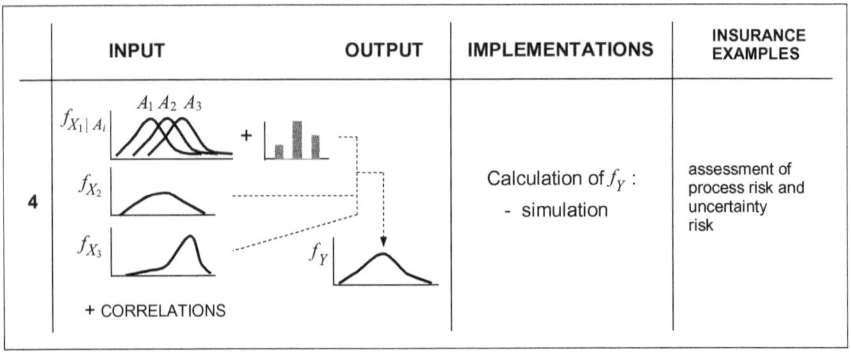

Fig. 4.20 From risk assessment to impact assessment: double-stochastic approach

Although the problem seems simple, finding the distribution of Y relying on analytical tools is a tricky business, especially according to the setting 2b, even if approximations are accepted. Hence, the implementation of these approaches must usually resort to stochastic simulation, which leads to the (conditional or unconditional) empirical distribution of Y and to the estimate of expected value, variance, etc.

Uncertainty about the distribution of the random variable X_1, for which three assumptions, A_1, A_2, A_3, have been adopted, is expressed in the setting sketched in Fig. 4.19. Three alternative conditional distributions of the random result Y are then derived. We note that this approach, although stochastic with respect to the random variables $X_1|A_1$, $X_1|A_2$, $X_1|A_3$, X_2, X_3, is quite deterministic in respect of the alternative assumptions,

The approach allowing for uncertainty, and hence systematic deviations, is progressed further in the setting represented in Fig. 4.20. A probability distribution on the space of assumptions is assigned; in other words, the three assumptions are weighted, according to the expert judgement. Hence, the unconditional distribution of Y can be calculated. The resulting setting can be denoted as double-stochastic approach. It is understood that the implementation of this setting calls for stochastic simulation.

Example 4.4 When a very simple setting is adopted, moving from risk assessment to impact assessment is a trivial step. Refer to Examples 4.1–4.3. The mortality law and the related parameters we have assumed constitute the only decision variable. Consider a portfolio of life annuities, initially consisting of N_0 annuitants all age x_0. The annual benefit b is paid to each annuitant at the end of each year as long as the annuitant is alive. Then, the total amount paid by the insurer at time t, $t = 1, 2, \ldots$, is simply given by:

$$B_t = b\, N_t$$

An example of deterministic assessment of impact on cash flows can immediately be derived from Example 4.1. Indeed:

$$\mathbb{E}[B_t] = b\, \mathbb{E}[N_t]$$

Further, a simple stochastic assessment can be obtained noting that:

$$\mathbb{P}[B_t = k\, b] = \mathbb{P}[N_t = k]; \quad k = 0, 1, \ldots, N_0$$

(see Eq. (4.6)) and then:

$$\mathbb{V}\mathrm{ar}[B_t] = b^2\, \mathbb{V}\mathrm{ar}[N_t]$$
$$\rho[B_t] = \rho[N_t]$$

Simulated paths of the cash flow streams $B_1, B_2, \ldots$ can simply be derived from the simulated paths of the numbers of survivors $N_1, N_2, \ldots$; see Example 4.2. Similarly, the impact of systematic deviations can be assessed starting from the idea underlying the approach adopted in Example 4.3 (and possibly defining a broad range of mortality scenarios). ∎

Remark A number of examples in Chaps. 6–8 will illustrate implementations of the approaches described above, more "realistic" than that described in Example 4.4.

4.4.5 Updating the Probabilistic Setting via Monitoring

As noted in Sect. 3.6 (see point 2), some aspects of the monitoring phase are strictly connected to data and hypotheses adopted to construct the model and hence to implement the step from risk assessment to impact assessment.

Adjusting data and hypotheses according to experience constitutes an important topic. Although this topic is beyond the scope of the present lecture notes, some basic ideas can help in understanding this feature of the monitoring phase.

We note what follows.

Fig. 4.21 Updating risk assessment and impact assessment via monitoring (I)

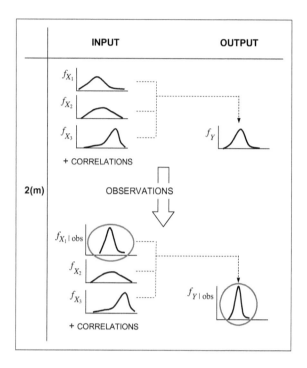

- The probability distributions of the scenario variables are assigned according to the expert's "initial" information (that is, the information available at a given time). So, for example, a probability distribution is assigned to the investment yield.
- The insurer's activity produces, thanks to scenario observations throughout time, new information.
- Updated information, and hence updated probability distributions can be constructed by combining the initial information with the new one. Appropriate statistical tools (e.g., Bayesian inference tools) can be used to this purpose.

The procedure sketched above is illustrated in Fig. 4.21, where it is assumed that the probability distribution of X_1, conditional on observations, is less dispersed thanks to the improved information state. A similar result then holds for the distribution of the output variable Y.

An updating procedure can also be applied to reduce the uncertainty expressed by a probability distribution over the space of assumptions, as illustrated by Fig. 4.22. According to this setting, the "prior" distribution is transformed into the "posterior" distribution, less dispersed thanks to the improved information state. A similar result then affects the output variable Y.

Fig. 4.22 Updating risk assessment and impact assessment via monitoring (II)

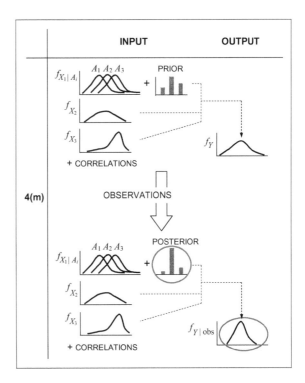

4.4.6 More on Deterministic Approaches

Deterministic approaches (briefly addressed at the beginning of Sect. 4.4.4), although not providing probabilistic information about the output variables, can play an important role in the impact assessment phase. Even though the relevant terminology is not univocally defined, the reader is referred to Fig. 4.23, where the objectives of deterministic approaches are sketched.

The expression *what-if analysis* generally denotes the calculation of the output values given the values assigned to scenario variables and / or decision variables. For example:

- what is the impact of a decrease in the interest rate adopted for premium calculation (thus, a decision variable)?
- what is the impact of a decrease in the estimated yield of investments (thus, a scenario variable) and a possible consequent decrease in the interest rate adopted for premium calculation (again, a decision variable)?

Assigning values to one or more scenario variables is commonly denoted as *scenario testing* (and sometime as *sensitivity testing*, when several alternative values are assigned to scenario variables). For instance:

- what is the impact of various assumptions on the surrender rates?

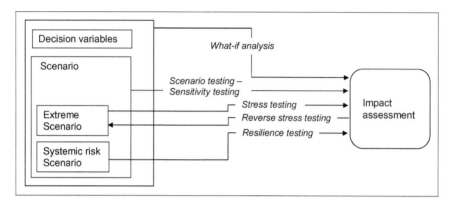

Fig. 4.23 Impact assessment via deterministic approaches

Stress testing denotes a particular scenario testing involving "extreme" values of (one or more) scenario variables. For example:

• what is the impact on an annuity portfolio results of an assumed permanent decrease of the annuitants' mortality?

We note that stress tests are adopted, for example, to determine solvency capital requirements to face adverse longevity scenarios (see Sects. 4.6.3 and 7.2.3).

A *reverse stress testing* aims at determining what set of values of a scenario variable (or more scenario variables) implies a critical situation in terms of a given output variable. For instance:

• what permanent decrease in the annuitants' mortality leads to a portfolio insolvency situation (e.g., in terms of insufficient portfolio reserves)?

Finally, *resilience testing* denotes a procedure aiming to assess the impact of a scenario affected by *systemic risk*, because of propagation of phenomena like the bankruptcy of several financial institutions (see Remark 2 in Sect. 4.1.2). The term "resilience" refers to the insurer's capacity to quickly recover from difficulties originated by the adverse scenario.

Remark *Profit testing* procedures were originally developed and implemented to perform deterministic assessments according to the logic of what-if analysis. The main purpose was to assess the impact of both decision variables (the premiums in particular) and scenario variables (for example the insureds' mortality, or the investment yield) on the annual cash flows of an insurance portfolio, and eventually on the profit generated by the portfolio itself (see also Sect. 3.7). Insurance language has recently attributed to the expression profit testing a much broader meaning, so encompassing stochastic valuations and assessment of the portfolio risk profile.

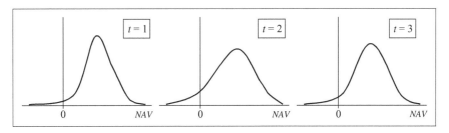

Fig. 4.24 The NAV distribution throughout time

4.5 RM Actions for Life Insurance

Important objectives of the RM process have been listed in Sect. 3.1, among which: profit, value creation, solvency, market share. Each RM action affects the achievement of the above objectives, and hence careful impact assessments must be performed. In this section, we address various RM actions for a life insurance portfolio, focusing in particular on the diversity of impacts.

4.5.1 Outlook

Among the quantities which represent the insurer's results, special attention is commonly placed on the NAV (Net Asset Value). This quantity, whose value will be denoted by NAV in what follows, is indeed referred to in solvency assessment schemes.

Future NAVs are, of course, random quantities. Assume that, by using an appropriate stochastic model (see, for example, the settings sketched in Fig. 4.18), we are able to "project" the NAV, that is, to calculate the probability distributions of NAV at times, say, $t = 1, 2, 3, \ldots$; see Fig. 4.24. We note that negative outcomes of NAV represent default situations. Further, positive but low values might represent critical situations from the solvency perspective.

RM actions must be taken in order to reduce the probability of low and, in particular, negative NAV outcomes. Three examples are given in Fig. 4.25. We note what follows.

- *Capital allocation* of course raises the expected value of the NAV distribution, which moves from E to E'.
- *Reinsurance*, while lowering the expected value of the NAV distribution from E to E'' because of the cost of the risk transfer, reduces the probability of poor outcomes of NAV.

Remark The choice of actions phase in the RM process usually results in a mix of actions (see Sect. 3.4), and this commonly occurs regarding capital allocation

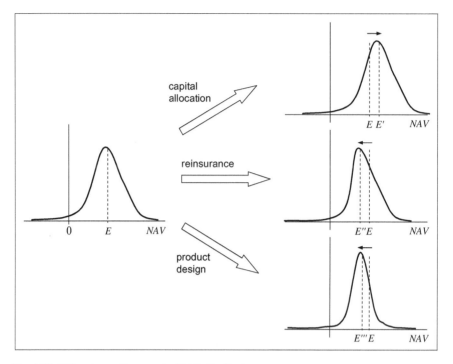

Fig. 4.25 RM actions: effects on the NAV distribution

and reinsurance. It is worth noting that reinsurance lowers the insurer's (expected) profit as a share of premiums must be ceded to the reinsurer; conversely, capital allocation, while contributing to the profit thanks to the investment yield, reduces the value creation because of the opportunity cost of the additional capital allocated (see Sect. 4.5.10).

- Thanks to the *product design*, the insurer can lower its risk exposure by reducing the range of guarantees and options included in the insurance products (see Chap. 5). As a result, on the one hand the NAV distribution is less dispersed but, on the other, a reduction of its expected value might follow.

4.5.2   Solvency

In line with common practice, we consider *solvency* to be the ability of the insurer to meet, with an assigned (high) probability, random liabilities as they are described by a realistic probabilistic structure. To implement such a concept, choices are needed in respect of the following items (see Fig. 4.15):

Fig. 4.26 Possible paths of the (portfolio) NAV

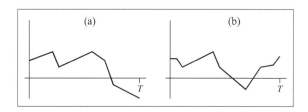

1. The quantity expressing the ability of the insurer to meet liabilities; a usual choice is given by the value of NAV (see Sect. 4.5.1).
2. The time span T which the above results are referred to; it may range from a short-medium term to the residual duration of the portfolio.
3. The timing of the results, in particular annual results (e.g., the value of NAV at every integer time within T years) versus single-figure results (e.g., the value of NAV at the end of the time horizon under consideration, that is, after T years).

As regards point 3, we note that, according to a single-figure approach, the NAV path in Fig. 4.26b fulfills the solvency requirement, while (depending on the timing of results) it might not fulfill the requirement if periodic results are taken into account. Hence, a single-figure approach can be adopted provided that borrowing money to fund the portfolio constitutes a possible solution. Of course, the NAV path in Fig. 4.26a does not fulfill the solvency requirement, whatever approach is adopted.

Finally, a choice concerns how to define the portfolio, that is, what policies are considered as belonging to the portfolio itself. Of course, the choice depends on the assessment purposes. For example, according to Solvency II Directive (Art. 101 (2)), the Solvency Capital Requirement "…shall cover existing business, as well as the new business expected to be written over the following 12 months".

Capital allocation as an action aiming at solvency is discussed in Sects. 4.5.9 and 4.5.10. Some formal aspects of solvency are addressed in Sect. 7.2 (in particular, see Examples 7.11 and 7.12), where a portfolio of life annuities is referred to.

4.5.3 Reinsurance: General Aspects

The reinsurance is the traditional risk transfer from an insurer (the *cedent*) to another insurer (the *reinsurer*). From a technical point of view, the basic aim of the reinsurance transfer is to find protection against the volatility, i.e. the risk of random fluctuations, in the portfolio payout, and eventually against the risk of the portfolio default (and the insurer's default, as well) as an extreme consequence of the volatility. Further aims of reinsurance will be addressed in Sect. 4.5.5.

The protection can be achieved via reinsurance arrangements which directly refer to the portfolio result, that is, *global reinsurance* arrangements. A commonly adopted alternative is provided by reinsurance arrangements which refer to each individual policy, i.e. *individual reinsurance* arrangements. See Fig. 4.27.

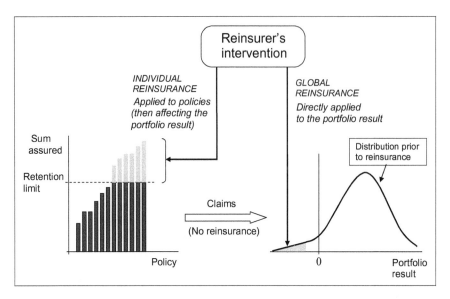

Fig. 4.27 Reinsurance arrangements: individual versus global

The rationale behind individual reinsurance arrangements can be found on the role of the distribution of sums assured in determining the impact of volatility on portfolio results (see Sect. 4.4.1). The lower the dispersion in the distribution of the sums assured, the better the diversification, the optimal portfolio consisting in policies with the same sum assured. Assuming the same portfolio size and the same average sum assured, Portfolio 1 in Fig. 4.28 has the lowest exposure to random fluctuations, whereas Portfolio 3 has the highest exposure. For a simple formal proof, the reader is referred to Olivieri and Pitacco (2015).

Reinsurance in life insurance, briefly *life reinsurance*, has specific features because of the characteristics of the life insurance business itself. We note in particular what follows:

- the amount of benefit is not random, and coincides with the sum assured (except for the possible presence of participation mechanisms which can increase the sum initially assured), unlike in non-life insurance where the benefit is usually defined as an indemnity or a reimbursement, and hence related to the severity of the claim;
- the number of events per policy (insured's death or policy maturity) which trigger the benefit is either zero or one, whereas in non-life insurance a policy can claim more times in a given period.

The above features simplify the definition of life reinsurance arrangements. Conversely, other problems arise because of the multi-year nature of life insurance policies. If the insurer is interested in purchasing protection against the mortality risk, the amount exposed to this risk, denoted as the *sum at risk* and given by the insured

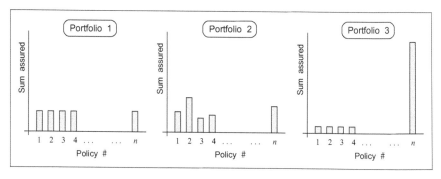

Fig. 4.28 Distribution of the sums insured in three portfolios

Fig. 4.29 Benefit, policy reserve and sum at risk in an endowment insurance policy

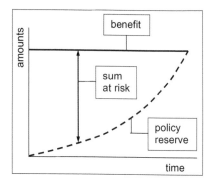

Fig. 4.30 Benefit, policy reserve and sum at risk in a term insurance policy

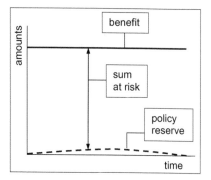

benefit less the policy reserve, is concerned. However, this amount varies throughout the policy duration because the annual change in the policy reserve.

For example, in an endowment insurance policy the reserve increases throughout the policy duration so that, assuming a constant benefit, the sum at risk decreases (see Fig. 4.29). Conversely, in a term insurance policy, the amount of the reserve is very low, so that the sum at risk is almost equal to the benefit throughout the whole policy duration (see Fig. 4.30).

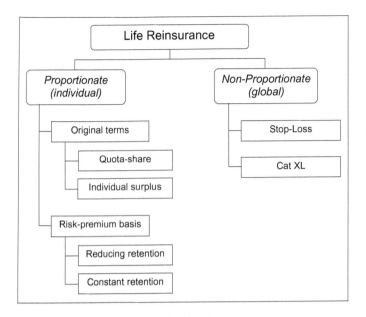

Fig. 4.31 Life reinsurance arrangements: a classification

4.5.4 Life Reinsurance Arrangements

Life reinsurance arrangements can be classified according to several criteria (e.g., global versus individual reinsurance arrangements, as already mentioned). We will follow the classification and the terminology proposed in Fig. 4.31.

In *proportionate reinsurance* arrangements, the amount ceded (or the sequence of amounts ceded throughout the policy duration) is stated at the time of the cession, and hence the reinsurer knows, before a claim occurs, the amount of its intervention. Conversely, in *non-proportionate reinsurance* arrangements, the amount of the reinsurer's intervention depends on the number and/or amounts of claims in a stated period. e.g., one year. In the life insurance context, non-proportionate reinsurance works at a portfolio level.

A proportionate reinsurance arrangement can be defined either in *original terms* or on a *risk-premium basis*. These are known in the US market as *coinsurance* and *yearly renewable term reinsurance*, respectively.

In original term reinsurance, according to a *quota-share* arrangement a fixed percentage, α, of each sum assured, is retained by the cedent, so that $1 - \alpha$ is transferred to the reinsurer. Denoting by C the generic sum assured, the retained amount is then given by:

$$C^{[\mathrm{ret}]} = \alpha\,C \tag{4.11}$$

Conversely, according to an *individual surplus* arrangement the part of the sum assured which exceeds the *retention limit* chosen by the cedent, $\bar{C}$, is transferred to

Fig. 4.32 Reinsurance in original terms: Quota-share

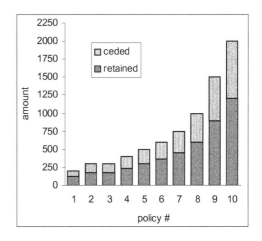

Fig. 4.33 Reinsurance in original terms: Individual surplus

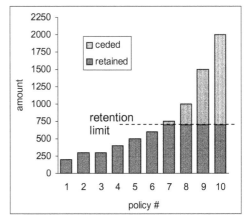

the reinsurer. The retained amount is then given by:

$$C^{[\text{ret}]} = \min\{\bar{C}, C\} \tag{4.12}$$

The two arrangements are illustrated in Figs. 4.32 and 4.33, respectively. We note that the surplus reinsurance has a leveling effect, in line with the target of reducing the dispersion in the distribution of sums assured and hence lowering the impact of random fluctuations.

Reinsurance on a risk-premium basis can be implemented in two different ways. First, refer to an endowment insurance policy (with annual level premiums). According to the *reducing retention method* (*RRM*) a stated percentage of the sum at risk is transferred to the reinsurer. Denoting by V_t the policy reserve at time t, the retained sum at risk is then given by:

$$(C - V_t)^{[\text{ret}]} = \alpha \, (C - V_t) \tag{4.13}$$

Fig. 4.34 Reinsurance on risk premium basis in an endowment insurance policy: RRM

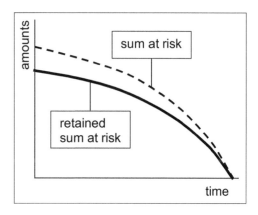

Fig. 4.35 Reinsurance on risk premium basis in an endowment insurance policy: CRM

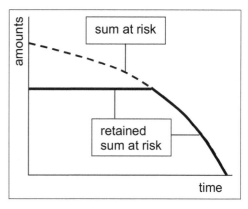

Conversely, according to the *constant retention method* (*CRM*), the excess (if any) of the sum at risk above a stated retention amount, $\bar{C}^{[R]}$, is transferred to the reinsurer. The retained sum at risk is thus given by:

$$(C - V_t)^{[\text{ret}]} = \min\{\bar{C}^{[R]}, C - V_t\} \tag{4.14}$$

Figures 4.34 and 4.35 illustrate the two retention methods. We note that, according to the RRM, the stated percentage (constant throughout the policy duration) implies a progressive reduction in both the amounts retained and ceded because of the decrease in the sum at risk in endowment insurance policies. Conversely, according to the CRM the amount retained remains constant as long as it is lower than the sum at risk.

As regards term assurance policies only providing a death benefit, the small amount of the reserve can be disregarded, and hence the sum at risk is assumed equal to the sum assured. Specific features of RRM and CRM then vanish.

Among the non-proportionate reinsurance arrangements (at a portfolio level), we focus on Stop-Loss reinsurance and Catastrophe reinsurance.

Stop-loss reinsurance provides a "direct" protection against the portfolio default, as it directly refers to the portfolio total payment. The reinsurer gets the reinsurance premium and pays the part of the portfolio total payment which exceeds a stated amount, the *stop-loss retention*, or *priority*. The priority is commonly expressed in terms of the total premium income (for example, $1.2\times$ premium income). An *upper limit* to reinsurer's intervention can be stated; in this case, the cedent remains exposed for the part exceeding the upper limit.

As the stop-loss reinsurance directly refers to the portfolio loss, it represents in theory the best solution to the portfolio protection. However, in practice, it should be noted that this reinsurance arrangement implies a potentially dangerous exposure of the reinsurer, related to the tail of the probability distribution of the portfolio payout (especially if no upper limit is stated). This means that a very high safety loading should be included into the reinsurance premium, possibly making this reinsurance cover extremely expensive. Hence, it is mainly used as an ingredient in a reinsurance programme, after other reinsurance covers (typically proportionate reinsurance covers, e.g., individual surplus) have been implemented to protect the portfolio.

The *Catastrophe reinsurance* (briefly, *Cat-XL*) aims to protect the portfolio (and the insurance company) against the risk that a single event (that is, a "catastrophe") causes a huge number of claims in the portfolio itself. For example:

- in a generic portfolio, a high number of claims might occur because of a disaster (hurricane, earthquake, pandemic, and so on);
- in a "group insurance", a number of insureds might die owing to a single accident in the workplace (explosion, fire, collapse, and so on).

A catastrophe is usually defined in terms of a given (minimum) number of claims, within a time interval of a given (maximum) duration, for example 48 hours. A deductible is usually stated, either in terms of number of claims or in monetary terms.

4.5.5 Reinsurance: Further Aspects

Reinsurance arrangements generally aim to keep the portfolio riskiness at a level acceptable by the ceding company. More reinsurance actually implies:

- a lower capital allocation;
- an increased underwriting capacity.

However, a more detailed analysis allows us to better understand the effects and the role of life reinsurance. In particular:

- a significant diversity of the effectiveness of the various reinsurance arrangements should be recognized;
- resorting to reinsurance can have purposes other than the reduction of the portfolio riskiness (due to insureds' mortality).

As regards the effectiveness of reinsurance arrangements, the following points should be considered.

1. Reinsurance in original terms implies a transfer of a proportion of all the policy components, i.e. the premium and the policy reserve. Transferring a proportion of the policy reserves leads to a loss in potential investment profits. For this reason, original term reinsurance is usually limited to term insurance policies, where reserves are very small and hence the loss in investment profit is not significant. Nevertheless, reinsurance in original terms can in general be useful for financing purposes, in particular thanks to the financing commission paid by the reinsurer to the cedent as remuneration for the placement of business with the reinsurer itself (for these reason, the expression *financial reinsurance* is sometimes used).

2. A surplus arrangement is more efficient than a quota-share, thanks to its "leveling" effect (see Fig. 4.33): it only aims at transferring part of policies with large amounts which significantly contribute to the portfolio riskiness. Conversely, a quota-share arrangement implies the transfer of a fixed proportion of all the business (see Fig. 4.32), hence reducing the premiums of small policies, which do not significantly contribute to portfolio riskiness, while implying a possibly dangerous retention of large policies.

3. Reinsurance transfer on risk-premium basis is more efficient than a transfer in original terms, as it allows to retain reserves in full, hence excluding transfer of reserves and consequent loss of potential investment profit.

4. Reinsurance arrangements mainly aim at reducing the impact of mortality random fluctuations and catastrophic events. In fact, the reinsurance company is willing to take the ceded risks as it can achieve an improved diversification of risks via pooling. Conversely, if the risk of systematic deviations heavily affects the portfolio, the reinsurer could reject the transfer, as these deviations concern the pool as an aggregate, and the total impact on portfolio results increases as the portfolio size increases. Notwithstanding, the reinsurer can take the risk of systematic deviations, with the proviso that a further transfer of this risk can be worked out. We will address this issue in Sects. 4.5.6 and 4.5.7. (Risk components and related impacts have been addressed in Sects. 4.1.2 and 4.1.3.)

Moving to the purposes of reinsurance, we note what follows.

- The cedent company can benefit from technical advice provided by the reinsurer. In particular:

  - the reinsurer, thanks to specific experience, can suggest statistical bases and inform about market features for new insurance products;
  - regarding in-force portfolios, the reinsurer can provide the cedent with an update of statistical bases (which is more effective if quota-share arrangements work, as they allow the reinsurer to monitor all claims pertaining to the reinsured portfolios).

- Reinsurance can have a "financing" role, in particular thanks to a sharing of policy and portfolio expenses between the cedent and the reinsurer (see also point 1 above).

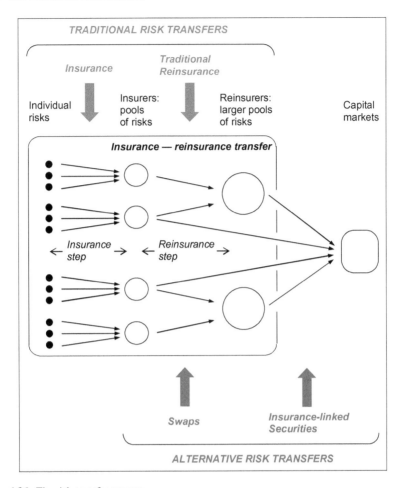

Fig. 4.36 The risk transfer process

4.5.6 The Risk Transfer Process

The risk transfer process consists of a sequence of steps (see Fig. 4.36). First, the *insurance–reinsurance transfer* can be split into two basic steps.

1. The *insurance step* consists in transferring risks from organizations (individuals, families, firms, institutions, and so on) to insurance companies; effects of this step are:

   a. building-up a pool;
   b. reducing the relative riskiness (caused by random fluctuations).

2. The *reinsurance step* consists in transferring risks from insurance companies (cedents) to reinsurers. This step can be implemented via traditional reinsurance arrangements, whose effects are:

   a. building-up larger pools;
   b. a further reduction of the relative riskiness (caused by random fluctuations).

   As an alternative, other arrangements can be adopted in the reinsurance step, e.g., swaps, belonging to the class of *Alternative Risk Transfers* (*ARTs*; see Sect. 4.5.7).

Traditional reinsurance works for the risk component caused by random fluctuations, which are diversifiable via pooling. However, risk components other than random fluctuations can affect insurers' and reinsurers' results, namely systematic deviations and catastrophic events. As regards the latter, larger pools can improve diversification, for instance thanks to an increased variety of geographical locations of insured risks. As regards the former, the relative impact of systematic deviations is independent of the pool size (and the absolute impact increases as the pool size increases). Thus, risk transfer arrangements other than traditional reinsurance, namely ARTs,

- are needed for transferring (at least to some extent) the risk of systematic deviations;
- can help in managing the catastrophe risk (lowering the cost of reinsurance, and/or the need for capital allocation).

ARTs intervene in the transfer from reinsurers to capital markets via *Insurance-linked Securities* (briefly *ILS*; see Sect. 4.5.7), but, as already noted, may also intervene in the transfer from insurers to reinsurers (e.g., via swaps).

In the following sections we will focus on ARTs in life insurance and reinsurance.

4.5.7   Alternative Risk Transfers. Securitization. ILS

The expression "Alternative Risk Transfers" commonly denotes a set of hedging tools other than the transfer via traditional reinsurance arrangements. Resorting to ARTs provides appropriate hedging solutions especially when transfer of risk components such as systematic deviations and catastrophe risks is needed (see also Sect. 4.1.3).

A (simplified) classification of ARTs in the context of life insurance and reinsurance, hence aiming at the transfer of biometric risks (mortality, longevity, and possibly disability) is sketched in Fig. 4.37. We note that two basic categories can be identified.

First, risks arising from contingent payments (benefits provided by insurance policies) can be packaged into securities traded on the capital market. The transaction is usually called *securitization*. Given the link of the pay-off of the securities (see below) with the insurer's (and/or reinsurer's) payments, the expression *insurance-linked securities* (briefly *ILS*) is commonly adopted. More specifically,

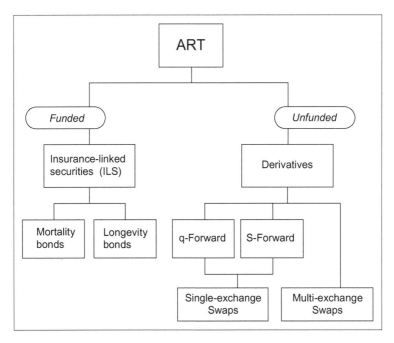

Fig. 4.37 Alternative Risk Transfers for biometric risks

when biometric risks are concerned, the expressions *mortality-linked securities* and *longevity-linked securities* are frequently used. Examples of ILS follow.

- *Mortality bonds* are used to (partially) transfer the risk of a mortality higher than expected, then implying an amount of death benefits paid by an insurer (or reinsurer) larger than expected. To this purpose, the issuer of the mortality bond (the insurer, or the reinsurer) pays reduced coupons and/or a reduced principal at maturity if the mortality in a given population, called the *reference population*, is higher than a stated benchmark, possibly owing to pandemics or natural disasters. Mortality bonds are typically short-term (3–5 years). More details are provided in Sect. 4.5.8.
- *Longevity bonds* aim at (partially) transferring the risk of a longevity higher than expected, hence implying an amount of survival benefits, typically life annuities, paid by an insurer (or pension fund, or reinsurer) larger than expected. Depending on its design, the longevity bond may offer hedging opportunities to a life annuity provider through either a long or a short position:

1. in the first case, the life annuity provider purchases longevity bonds, whose payoff increases as mortality decreases;
2. in the second case, longevity bonds are issued by the annuity provider or the reinsurer and sold to investors; in this case the bond payoff decreases as mortality decreases.

The longevity bonds are typically long-term securities (20 or more years), because:

– the longevity risk reveals over a long period of time;
– the annuity provider needs to offset benefit payments throughout long durations, as life annuities are payable lifelong.

Given the long term maturity, in any case it is reasonable that the link is realized through the coupon, hence providing hedging on a yearly basis. More details can be found in Sect. 4.6.7.

In the framework of ILS, the following securities can also be placed:

• *cat-bonds*, for transferring the risk of huge benefit payments due to some catastrophic event (earthquake, flood, etc.).

Alternative Risk Transfers belonging to this category are called *funded ARTs*, as the relevant transactions are based on securities. Investing in ILS (e.g., in longevity bonds of type 2) basically relies on a diversification target, assuming that the yield provided by an ILS is (reasonably) uncorrelated with the performance of most of other securities traded on the capital market. However, a counterparty risk arises for the investor, because of possible default of the bond issuer.

We note that ILS issued by a risk owner and sold to investors implement a hedging strategy denoted as approach 1(b) in Sect. 3.4.2. Indeed, the higher is the benefit payout, the smaller is the bond payoff (either in terms of coupons, or principal at maturity, or both).

The experienced mortality (or longevity) which is compared to the agreed benchmark is usually the one observed in the reference population, and not in the specific insurance portfolio for which the hedging strategy is implemented. Then, a *basis risk* arises, because of possible imperfect hedging due to different age-patterns of mortality in the population and the portfolio respectively.

When a reference population is considered in defining the ART, the risk transfer is denoted as *index-based* (or *standardized*), as the population mortality is usually expressed by an appropriate index. Conversely, in the case the actual portfolio mortality is compared to the benchmark mortality, the risk transfer is called *indemnity-based* (or *customized*). Clearly, an index-based transfer is preferred by investors, as population mortality data are collected and the index calculated by independent analysts.

The same argument, regarding possible imperfect hedging, also applies to the derivatives described below.

Remark Motivations other than a risk transfer can underly a securitization transaction. A securitization can consists in packaging a pool of assets (in particular intangible assets) or, more generally, a cash flow stream into securities traded on the capital market. The aim of such a securitization transaction is to raise liquidity by selling future flows. In the insurance and reinsurance context, the specific aim can be the recovery of acquisition costs (especially in life insurance) or expected profits.

Secondly, specific derivatives, with mortality (or longevity, or disability) in a given population as the underlying, can be used to face the biometric risks. Examples are as follows.

- The *q-forward* (the letter q usually denotes a probability of dying) is a contract according to which an amount linked to the observed mortality rate in the reference population at a given future date (the maturity of the contract) will be exchanged at maturity in return for an amount linked to a benchmark mortality rate agreed at the time the contract is written.
- The *S-forward* (the letter S usually denotes the survival function, and hence a probability of being alive) is a contract according to which an amount linked to the observed survival rate in the reference population at a given future date (the maturity) will be exchanged at maturity in return for an amount linked to a benchmark survival rate agreed at the time the contract is written.

We note that the q-forward and the S-forward realize a *single-exchange swap*.

- In general terms, a *(multi-exchange) swap* is a derivative according to which two counterparties periodically exchange cash flows. If the underlying is the mortality (or longevity) in the reference population, the swap can be thought as a sequence of q-forwards or S-forwards in which all the benchmark mortality (or longevity) rates are stated at the time the swap contract is written.

Alternative Risk Transfers belonging to this category are called *unfunded ARTs*, as no security is issued.

4.5.8 Mortality Bonds

Mortality bonds belong to the class of ILS. The structure of a risk transfer via mortality bonds, which involves a sequence of transactions, is sketched in Fig. 4.38.

1. The following transactions take place in the insurance/reinsurance market.

    a. Insurers issue policies and cash premiums from policyholders, then pay out benefits.
    b. Insurers cede to a reinsurer shares of the risks taken and pay reinsurance premiums; the reinsurer pays shares of benefits to insurers.

2. The following transactions involve, in particular, the reinsurer and the capital market, i.e. the investors.

    a. The reinsurer enters into an agreement with a *Special Purpose Vehicle (SPV)*, that is, a specific entity used for the transfer deal. Using an SPV has important advantages for both the reinsurer (cash flows inherent the mortality bond are kept off the reinsurer's balance) and the investors (credit risk is reduced).
    b. According to the agreement, the SPV will made floating payments to the reinsurer if a certain *mortality index* is triggered. Against the above payments, the

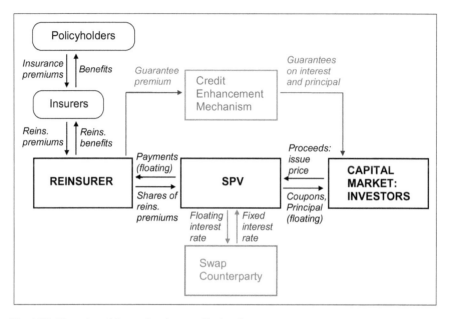

Fig. 4.38 Flows in a risk transfer via mortality bond

reinsurer periodically pays fixed amounts, given by shares of the reinsurance premiums.

c. The SPV raises funds to maintain the floating payments to the reinsurer, by issuing and selling a mortality bond. Hence, the SPV:

- cashes the proceeds from the bond sale and invests in high-quality securities (which act as a collateral);
- pays to the investors coupons and principal, whose amounts can be reduced according to the behavior of the mortality index (formal definitions of the mortality index and the link between mortality index and bond payoff are given in Examples 4.5, 4.6 and 4.7).

3. Further entities can be involved in the risk transfer via mortality bond.

a. The intervention of a Swap counter-party is required if a fixed interest rate is the baseline reward to investors.

b. To reduce default risks, some form of credit enhancement may be introduced. In particular, a specific entity (issuing, for example, credit insurance, letters of credit, and so on) can be involved. Note that intervention by a specific financial institution may anyhow result in an increase in the rating of the securities.

In the examples: 0 is the time of issue of the mortality bond, and T its maturity. The following notation is adopted:

- V_t denotes the value of the principal of the bond at time t, for $t = 0, 1, \ldots, T$;
- C_t is the coupon due at time t, for $t = 1, 2, \ldots, T$;
- $I(t)$ denotes the mortality index at time t, for $t = 0, 1, \ldots, T$.

Example 4.5 An example of mortality index is as follows:

$$I(t) = q^{[\text{obs}]}(t); \quad t = 0, 1, \ldots, T \tag{4.15}$$

where $q^{[\text{obs}]}(t)$ is the observed annual mortality rate averaged over the reference population in year t. Although just some ages could be considered in detecting situations of high mortality, it is reasonable to address a broad range of ages. To summarize the mortality experienced over the whole lifetime of the bond, the following indexes can be adopted:

$$I_T = \max_{t=1,2,\ldots,T} \{I(t)\} \tag{4.16a}$$

$$I_T = \frac{\sum_{t=1}^{T} I(t)}{T} \tag{4.16b}$$

We note that, according to both the definitions, I_T constitutes a marker of the mortality experienced over the T-year interval.   ∎

Example 4.6 Assume that the bond aims at protecting against high mortality experienced over the whole lifetime of the bond itself. Hence, index (4.16a) or (4.16b) should be adopted. Hedging is obtained by reducing the principal at maturity. Hence, the principal paid-back to investors can be defined, for example, as follows:

$$V_T = V_0 \times \begin{cases} 1 & \text{if } I_T \leq \lambda' I(0) \\ \varphi(I_T) & \text{if } \lambda' I(0) < I_T \leq \lambda'' I(0) \\ 0 & \text{if } I_T > \lambda'' I(0) \end{cases} \tag{4.17}$$

where λ' and λ'' ($1 \leq \lambda' < \lambda''$) are two parameters (stated in the bond conditions), and $\varphi(I_T)$ is a proper decreasing function, such that $\varphi(\lambda' I(0)) = 1$ and $\varphi(\lambda'' I(0)) = 0$. For example:

$$\varphi(I_T) = \frac{\lambda'' I(0) - I_T}{(\lambda'' - \lambda') I(0)} \tag{4.18}$$

See Fig. 4.39. The coupon is independent of the experienced mortality. In particular, it can be given by:

$$C_t = V_0 (i_t + r) \tag{4.19}$$

where i_t is the market interest rate at time t, and r is an extra-yield rewarding investors for taking the mortality risk.   ∎

Fig. 4.39 Mortality bond: principal payout

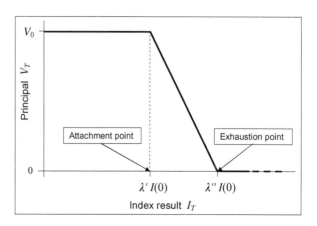

The cash flows related to the bond described in Example 4.6 aim to match the flows in the life insurance portfolio just at the end of a period of T years. An alternative design of the mortality bond can be conceived to provide a match on a yearly basis.

Example 4.7 To provide a match on a yearly basis, the mortality index $I(t)$ (e.g., defined according to (4.15)) should be used. Then, assume that the coupon is given by:

$$C_t = V_0 \times \begin{cases} i_t + r & \text{if } I(t) \le \delta' q(t) \\ (i_t + r) \psi(I(t)) & \text{if } \delta' q(t) < I(t) \le \delta'' q(t) \\ 0 & \text{if } I(t) > \delta'' q(t) \end{cases} \qquad (4.20)$$

where $q(t)$ is the mortality rate in the reference population according to a given mortality assumption, and δ', δ'' are two parameters (stated in the bond conditions), with $1 \le \delta' < \delta''$. The function $\psi(I(t))$ must then be decreasing, with $\psi(\delta' q(t)) = 1$ and $\psi(\delta'' q(t)) = 0$; for example:

$$\psi(I(t)) = \frac{\delta'' q(t) - I(t)}{\delta'' q(t) - \delta' q(t)} \qquad (4.21)$$

As in (4.19), the rate r in (4.20) is the extra-yield rewarding investors for the mortality risk inherent in the bond payoff. Note that, in this structure, the principal at maturity can be assumed independent of the experienced mortality, for example:

$$V_T = V_0 \qquad (4.22)$$

∎

The following features of a mortality bond should finally be noted.

- As mortality bonds mainly aim at hedging extreme mortality, usually occurring in a short time frame, a short duration (say, 3–5 years) is appropriate. It follows that

Fig. 4.40 Assets, Liabilities and Shareholders capital

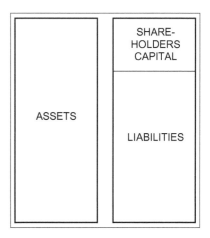

hedging can be realized via reduction of the principal only, as shown in Example 4.6, hence without affecting the coupons.

- The index adopted to determine the bond payoff (coupons or principal) expresses the mortality experienced in the reference population, not necessarily coinciding with the insureds' mortality. Then, because of the basis risk (see Sect. 4.5.7), a perfect hedging is not guaranteed.

4.5.9 *Capital Allocation*

Insurance activity implies insurer's obligations towards the policyholders, i.e. liabilities. Of course, liabilities are random and deferred in time. Insurer's liabilities are counterbalanced by assets.

Because of randomness in insurance liabilities, the amount of assets must be greater than the expected value of the liabilities (assessed on a realistic basis), in order to guarantee the insurer's capability to fulfil, with a high probability, its obligations.

A generic balance-sheet structure is shown in Fig. 4.40. The following points clearly emerge:

1. specific rules must be adopted to evaluate assets and liabilities;
2. a specific criterion must be adopted to assess the insurer's capability to fulfil its obligations;
3. choices 1 and 2 above will result in the need for allocating to the insurer's portfolio(s) a part of the shareholders' capital, in order to guarantee the above capability.

Remark 4.1 In what follows, we basically refer to the current European Union solvency regulation. Diverse principles of reserving must be adopted according to IFRS

(International Financial Reporting Standards) and local GAAP (Generally Accepted Accounting Principles). See also Remark 4.2 in this section.

Assume that the assets are assessed at their market value, which can be interpreted as their "true" or "fair" value. Then, also the related liabilities should be assessed, for consistency, at market value. Thus, the so-called *mark-to-market* approach to the assessment of both assets and liabilities should in principle be adopted. However, a problem arises: is a (reliable) market value of liabilities available?

As insurers' liabilities are only traded in markets which cannot provide a reliable fair value (for example, the reinsurance market), the application of the mark-to-market approach is restricted to liabilities which can perfectly be hedged by assets traded on appropriate markets. This is the case, in particular, of the liabilities related to unit-linked insurance products.

As a practicable alternative, the so-called *mark-to-model* approach to the assessment of the portfolio liabilities can be implemented. This approach relies on an actuarial model whose output should provide a reasonably fair value of the liabilities.

The fair value of liabilities then consists of the two following components:

- the *best-estimate reserve* (or *BEL*, that is, *best-estimate liability*), calculated as the expected present value (that is, the actuarial value) of the future portfolio outflows; a "realistic" probability structure (mortality, probability of disablement, probability of exercising contractual options, etc.) is adopted;
- the *risk margin*, which is defined as the cost (beyond the risk-free rate) of the solvency capital (see below) which is required for the run-off of the portfolio in the case of insurer's default at the end of the current year, assuming that another insurer is charged with the portfolio itself. Hence:

  – the risk margin makes possible the run-off of the portfolio after default;
  – without risk margin, no other insurer would be available to be charged with the portfolio itself;
  – the risk margin "belongs" to the policyholders, because in the case of default it must be transferred together with the portfolio; thus, it is not a part of the shareholders' capital.

Remark 4.2 As noted in Remark 4.1, other criteria can (or must) be adopted in the assessment of liabilities. In particular, a prudent (or safe-side) reserve quantifies the liabilities according to various GAAP, instead of a best-estimate reserve plus a risk margin.

More assets than those just backing the fair value of the liabilities are usually needed to face risks inherent in liabilities. To this purpose, shareholders' capital must be allocated and assigned to the portfolio (see point 3 above). The amount to be allocated to a portfolio (and, more in general, to a life insurance business) is determined according to a stated *solvency* target. Thus, the total amount of assets backing the insurer's liabilities (assessed in terms of fair value) and the allocated shareholders' capital must fulfill the *adequacy requirement*, as stated by the supervisory authorities

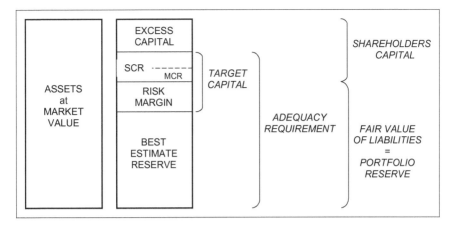

Fig. 4.41 Assets, Liabilities and Shareholders capital, according to specific evaluation and solvency criteria

(or by the company management, if the latter results in an amount higher than that required by the authorities).

The shareholders capital needed to fulfill the adequacy requirement is determined according to the *solvency capital requirement* and usually denoted as the *SCR*. The (possibly) remaining shareholders' capital constitutes the *excess capital*. The so called *target capital* consists of two components, namely the risk margin (a part of the portfolio reserve) and the SCR (a part of the shareholders' capital). Hence, the adequacy requirement is fulfilled by: (1) the best-estimate reserve, (2) the risk margin, and (3) the SCR. See Fig. 4.41.

SCR must be calculated either according to a *standard formula*, or an internal model. In addition to SCR, a *minimum capital requirement (MCR)* must be calculated, whose amount represents the threshold below which the national supervisory authority would intervene.

4.5.10 Solvency Versus Value Creation

The higher the amount of capital allocated to the portfolio, the higher is the solvency level (and hence, the lower the probability of default). However, capital is a limited resource and, whatever the amount allocated, the capital allocation has a cost, the so called *opportunity cost*.

The cost of capital is not an item of the profit-and-loss statement, but constitutes a (negative) component when determining the value creation. Indeed, the value creation is defined as the (positive) difference between the revenues and the costs associated to all of the production factors, hence including the cost of the capital invested in the business. In this sense, value creation is a synonym for (positive) *economic earnings*.

Remark As is well known, various meanings can be attributed to the word "value" and hence to the expression "value creation". Given the above definition, we are here referring to value creation from the shareholders' perspective.

Solvency and value creation are two objectives, or targets, which should drive the RM process (see Sect. 3.1). However, the conflict between the two objectives is evident.

Capital allocation is a key item in any business model. Although a detailed discussion on this topic is beyond the scope of this presentation, a very simple formal model can help in understanding the interaction between the combined result of some basic RM actions and value creation.

For simplicity, we refer to a one-year insurance portfolio (or a one-year "segment" of a multi-year portfolio, e.g., of term insurance policies), whose random payout is denoted by X. Let Π denote the total amount of premiums (one-year "natural" premiums in the case of multi-year policies) facing the benefit payment X, and assume that Π is given by:

$$\Pi = P + m \tag{4.23}$$

where:

- P is the total amount of equivalence premiums, that is, $P = \mathbb{E}[X]$ (according to the realistic basis, and disregarding the time-value of money over the year);
- m denotes the total safety loading, equal, at the same time, to the expected value of the portfolio random result $Z = \Pi - X$, i.e. the (expected) *profit margin*:

$$\mathbb{E}[Z] = \mathbb{E}[\Pi - X] = \Pi - P = m \tag{4.24}$$

We assume that M is the amount of capital allocated to the portfolio. The event $[X > \Pi + M]$ then denotes the portfolio *default*, and hence an *insolvency* situation. The default probability is given by:

$$\mathbb{P}[X > \Pi + M] = \mathbb{P}[X > P + m + M] = \mathbb{P}\left[\frac{X - P}{\sigma} > \frac{m + M}{\sigma}\right] \tag{4.25}$$

where σ denotes the standard deviation of the payout X.

We focus on the unit-free index s,

$$s = \frac{m + M}{\sigma} \tag{4.26}$$

We note that the higher is s, the lower is the default probability. To raise s, the following RM actions can be taken:

1. raise the safety loading m;
2. raise the capital allocation M;
3. reduce σ, via appropriate reinsurance arrangements (thus affecting the portfolio structure, in terms of sums assured), and, in particular, by choosing the retention limit (see Sects. 4.5.3 and 4.5.4).

Amount of safety loading, capital allocation and retention level constitute decision variables, according to the terminology defined in Sect. 4.4.3. However, the following aspects should be stressed.

- For any given σ, the probability of default only depends on the sum $m + M$. Nonetheless, actions involving either one or the other of the two terms imply different effects (see the following points).
- Action 1 affects the premiums. The related cost is charged to the policyholders, and hence the action is bounded by market constraints (in the present setting, we must assume that the action does not reduce the market share and hence the total payout).
- Action 2, whose cost is charged to the shareholders, has constraints at the company level, because capital is a limited resource and, anyway, the cost of capital allocation affects value creation.
- As regards action 3, whatever reinsurance arrangement may be chosen, the related cost obviously affects the resources available to the portfolio, in particular reducing the expected profit m. As both numerator and denominator of the index s are affected (see Eq. (4.26)), the effect is not univocally determined in general.

Disregarding action 3, we now focus on the combined effect of actions 1 and 2 on the value creation. In terms of expected values, the profit margin m must be compared to the cost of capital, which is given by $r\,M$, with r denoting the opportunity-cost rate. Hence:

$$m > r\,M \quad \Leftrightarrow \quad \text{value creation}$$
$$m = r\,M \quad \Leftrightarrow \quad \text{no value creation}$$
$$m < r\,M \quad \Leftrightarrow \quad \text{value destruction}$$

The combined effects of RM actions 1 and 2 in terms of both value creation and probability of default (and hence solvency) are shown in Fig. 4.42.

4.6 RM Actions for Life Annuities

As noted in Sect. 4.4.2, a number of portfolio results can be taken as "metrics" to assess the effectiveness of RM actions. In what follows, we only focus on annual cash flows pertaining to annuity payments, that anyhow constitute the starting point from which other quantities (e.g., profits) can be derived.

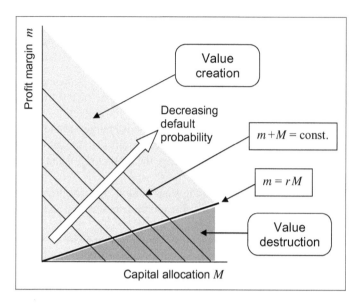

Fig. 4.42 Capital allocation: value creation versus value destruction

4.6.1 Expected Versus Actual Cash Flows

In Fig. 4.43, which refers to a single-premium life annuity portfolio consisting of one cohort, a sequence of possible outcomes of the annual cash flows is represented (the bars), together with the profile of the expected values on a realistic basis of the random cash flows (the green line), and a barrier (the black line), that is the "threshold" which represents a maintainable level of benefit payments. The threshold amount is financed by premiums via the portfolio reserve (including the risk margin; see Sect. 4.5.9), and by shareholders' capital as the result of the allocation policy (consisting of specific capital allocations as well as accumulation of undistributed profits). The situation depicted in Fig. 4.43, where some annual outflows exceed the threshold level, should clearly be avoided. To lower the probability of such critical situations, the insurer can resort to various RM actions.

A number of RM actions are displayed in Fig. 4.44, which aim at risk mitigation, meant as lowering the frequency and the severity of events like those which imply the situation depicted in Fig. 4.43. In practical terms, a RM action can have the following targets:

- an uplift of the maintainable annual outflows, thus a higher threshold level;
- lower (and smoother) annual outflows in the case of unanticipated improvements in portfolio mortality.

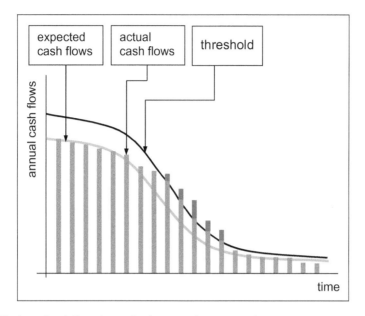

Fig. 4.43 Annual cash flows (one cohort): expected versus actual amounts

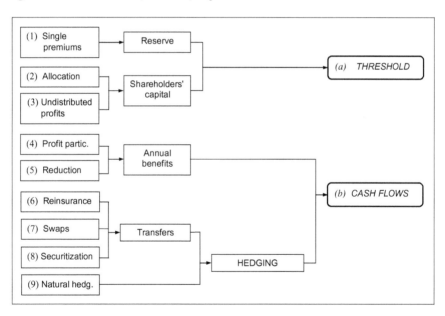

Fig. 4.44 RM actions for a life annuity portfolio

4.6.2   Pricing and Product Design

The pricing of life annuity products constitutes a tool for limiting possible losses. This action is represented by path $(1) \rightarrow (a)$ in Fig. 4.44. The following issues should in particular be taken into account.

- Mortality improvements require the use of a projected life table for pricing life annuities.
- Because of uncertainty in future mortality trend, a premium principle other than the traditional equivalence principle should be adopted. It is worth noting that, adopting the equivalence principle, the longevity risk can be accounted for only via a (rough) safety loading, constructed by increasing the survival probabilities resulting from the projected life table. Indeed, this approach is often adopted in the current actuarial practice.
- The presence, in an accumulation product such as an endowment, of an option to annuitise at a fixed annuitization rate (the so-called *Guaranteed Annuity Option*, briefly the *GAO*) requires an accurate pricing model accounting for the value of the option itself.

To limit the amount of possible losses, it is necessary to control the annuity amounts paid out. Hence, some flexibility should be added to the life annuity product, aiming to weaken some of the guarantees (see Sect. 5.5). One action could be the reduction of the annual amount as a consequence of an unanticipated mortality improvement (path $(5) \rightarrow (b)$ in Fig. 4.44). However, in this case the product would be a non-guaranteed life annuity (though with a reasonable minimum amount guaranteed). A more practicable tool, consistent with the features of a guaranteed life annuity, is the reduction of the level of investment profit participation while a poor mortality is experienced (path $(4) \rightarrow (b)$); it is worth stressing that undistributed profits also increase the shareholders' capital within the portfolio, hence uplifting the maintainable threshold (path $(3) \rightarrow (a)$). See also Sect. 5.5.

4.6.3   Capital Allocation

General issues on capital allocation have been presented in Sect. 4.5.9. Hence, we here focus on specific issues regarding the life annuity business.

A life annuity portfolio is of course exposed to the longevity risk. In particular, the aggregate longevity risk (that is, the risk of systematic deviations) cannot be diversified via pooling, and hence an appropriate capital allocation is required (path $(2) \rightarrow (a)$ in Fig. 4.44).

The solvency capital requirement can be determined via stress testing (see Sect. 4.4.6). A stress scenario can be defined, for example, in terms of:

1. a given increase in the life expectancy at, say, age 65;
2. a given decrease in the annual probabilities of death from, say, age 65 onwards (which, of course, implies a raise in life expectancy).

Approach 2 has been adopted, for example, by the Solvency II Directive in the European Union: an amount of capital must be allocated, such that the insurer can meet the payment of annuity benefits in the case a permanent decrease of 20% in the annual probabilities of death occurs. A formal definition of this approach is given in Sect. 7.2.3.

4.6.4 Hedging and Risk Transfers: General Issues

While pricing and capital allocation aim at raising the threshold, profit participation mechanisms and possible benefit reduction affect the amounts of annual cash flows. Other RM actions also affect the cash flows via various hedging strategies: risk transfers and natural hedging.

Risk transfer solutions must in general be chosen looking at the importance of the diverse risk components and related impacts (see Sects. 4.1.2 and 4.1.3). Of course, both the risk of random fluctuations and the risk of systematic deviations impact on portfolio results.

The risk of mortality random fluctuations can be diversified via pooling, and hence can be transferred to a reinsurer which, thanks to a larger portfolio size, will be able to achieve a better diversification. Conversely, the risk of systematic deviations cannot be diversified inside the traditional insurance-reinsurance transfer. Hence, as the two risk components affect the annuity portfolio at the same time, to gain effectiveness reinsurance transfers must be completed with a further transfer. Thus, the traditional reinsurance transfer can be accepted provided that the reinsurer has access to further arrangements through which the systematic risk component can also be transferred.

Diverse risk transfer solutions are sketched in Fig. 4.44: (traditional) reinsurance arrangements (path $(6) \rightarrow (b)$), swap-like reinsurance $((7) \rightarrow (b))$ and securitization via specific financial instruments, e.g., longevity bonds $((8) \rightarrow (b))$, whose performance is linked to some measure of longevity in a given population.

The above solutions, as well as natural hedging possibilities, are described in the following sections.

4.6.5 Reinsurance

As is well known, a homogeneous portfolio (in terms of benefit amounts) is less risky than a heterogeneous portfolio (see Sect. 4.5.3). Hence, a traditional *Surplus reinsurance* arrangement can improve the diversification via pooling. See Fig. 4.45.

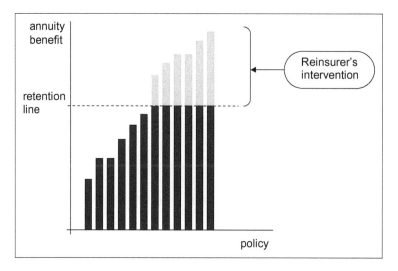

Fig. 4.45 A Surplus reinsurance arrangement

A natural way to transfer longevity risk for an annuity provider is to truncate the duration of each annuity. To this purpose, an (individual) *Excess-of-Loss (XL)* reinsurance can be designed. Under such arrangement, the reinsurer would pay to the cedent the "final" part of the life annuity while exceeding a given age $\bar{x}$. Such an age should be reasonably old, but not too close to the maximum attainable age (otherwise the transfer would be ineffective); $\bar{x}$ could, for example, be set equal to the modal age of death (the so called Lexis point) in the current projected life table. Note that $\bar{x}$ determines the deductible of the XL arrangement. See Fig. 4.46, where $\bar{x} = 85$.

From the point of view of the cedent, this reinsurance arrangement converts immediate life annuities payable for the whole residual lifetime into immediate temporary life annuities. From the point of view of the reinsurer, a severe charge of risk emerges. Actually, the reinsurer takes the "worst part" of each annuity, being involved at the oldest ages only, that is, on the tail of the lifetime distribution. Therefore from a practical point of view, the treaty is acceptable by the reinsurer only if it is compulsory for some annuity providers. This could be the case, for example, of pension funds, which may be forced by the supervisory authority to back their liabilities through arrangements with (re-)insurers.

A *Stop-loss reinsurance* may be designed on annual outflows. The rationale is that, at time t, longevity risk is perceived if the amount of benefits, B_t, to be currently paid to annuitants is (significantly) higher than expected. A transfer arrangement can then be designed so that the reinsurer takes charge of such extra-amount, or "loss". As in the previous cases, the loss may be due to random fluctuations only; setting a trigger level for the reinsurer's intervention higher than the expected value of the amount of benefits reduces the possible transfer of such risk component, and hence

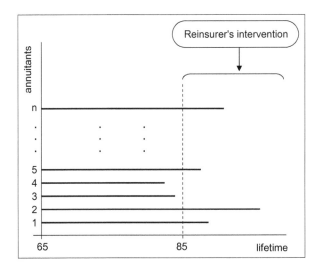

Fig. 4.46 An XL reinsurance arrangement

the cost of the reinsurance arrangement. Reinsurance conditions should concern the following items.

- Assume, for simplicity, that the reinsurance agreement is written at the beginning of payments to a cohort of annuitants. The time horizon k of the reinsurance coverage must be stated, as well as the timing of possible reinsurer's intervention within it. Within the time horizon k, in particular, policy conditions (i.e. premium basis, mortality assumptions, and so on) should be guaranteed. As to the timing of the reinsurer's intervention, since reference is to annual outflows, it is reasonable to assume that a yearly timing is chosen; in the following, we will refer to this arrangement.
- A mortality scenario, S, must be assumed (e.g., the best-estimate scenario), to calculate the expected value of the cash flows, $\mathbb{E}[B_t \mid S]$, required to define the loss of the cedent.
- The amount Λ'_t of benefits at time t ($t = 1, 2, \ldots, k$) below which there is no payment by the reinsurer must be stated. For example:

$$\Lambda'_t = \mathbb{E}[B_t \mid S] (1 + \alpha) \tag{4.27}$$

with $\alpha \geq 0$; thus, the amount Λ'_t represents the priority of the Stop-Loss arrangement.
- The Stop-Loss upper limit must be defined, i.e. an amount Λ''_t, such that $\Lambda''_t - \Lambda'_t$ is the maximum amount paid by the reinsurer at time t. From the point of view of the cedent, the amount Λ''_t should be high enough to let charged to the cedent just situations of extremely high survivorship. However, the reinsurer reasonably sets Λ''_t according to the available hedging opportunities. As to the cedent, a further

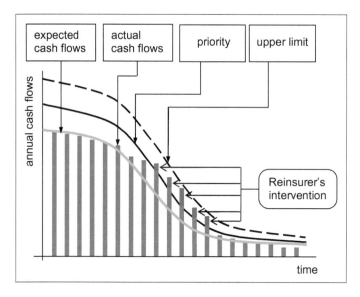

Fig. 4.47 A Stop-Loss reinsurance arrangements on annual outflows

reinsurance arrangement may be underwritten, if available, for the residual risk, possibly with another reinsurer; in this case, the amount $\Lambda_t'' - \Lambda_t'$ works as the first layer.

An example of priority, upper limit and reinsurer's intervention is given by Fig. 4.47. The flow, $B_t^{[SL]}$, paid by the reinsurer at time t ($t = 1, 2, \ldots, k$) is then given by:

$$B_t^{[SL]} = \begin{cases} 0 & \text{if } B_t \leq \Lambda_t' \\ B_t - \Lambda_t' & \text{if } \Lambda_t' < B_t \leq \Lambda_t'' \\ \Lambda_t'' - \Lambda_t' & \text{if } B_t > \Lambda_t'' \end{cases} \tag{4.28}$$

The net outflow of the cedent at time t (gross of the reinsurance premium), denoted by $OF_t^{[SL]}$, is then:

$$OF_t^{[SL]} = B_t - B_t^{[SL]} = \begin{cases} B_t & \text{if } B_t \leq \Lambda_t' \\ \Lambda_t' & \text{if } \Lambda_t' < B_t \leq \Lambda_t'' \\ B_t - (\Lambda_t'' - \Lambda_t') & \text{if } B_t > \Lambda_t'' \end{cases} \tag{4.29}$$

The net outcome of the cedent is still random but, unless some "extreme" survivorship event occurs, protected with a cap. We note that the effect of the Stop-Loss arrangement is to transfer to the reinsurer all the loss situations, except for the poorest and the heaviest ones; any situation of profit, on the contrary, is kept by the cedent.

4.6.6 Longevity Swaps

To reduce randomness of the annual outflows, the cedent may be willing to transfer to the reinsurer not only losses, but also profits. To this purpose, a *reinsurance swap arrangement* (also called *longevity swap*, or *survivor swap*) on annual outflows can be designed. See Sect. 3.4.2 for an introduction to the swap rationale.

Let B_t^* denote the target value for the insurer's outflow at time t; for example:

$$B_t^* = \mathbb{E}[B_t \mid S] \tag{4.30}$$

where S denotes a mortality scenario (e.g., the best-estimate scenario). Under the swap agreement:

- if $B_t > B_t^*$, then the cedent receives money from the reinsurer;
- if $B_t < B_t^*$, then the cedent pays money to the reinsurer.

The amount paid (received) by the reinsurer at time t is then:

$$B_t^{[\text{swap}]} = B_t - B_t^* \tag{4.31}$$

so that the annual outflow (gross of the reinsurance premium) for the cedent is given by:

$$OF_t^{[\text{swap}]} = B_t - B_t^{[\text{swap}]} = B_t^* \tag{4.32}$$

The advantage for the cedent is to convert a random flow, B_t, into a flow certain, B_t^* (whence the term reinsurance swap assigned to this arrangement). A possible situation is sketched in Fig. 4.48. We note that, *ceteris paribus*, this arrangement should be less expensive than the Stop-Loss on outflows, given that the reinsurer participate not only in losses, but also in profits.

Although an advantage for the cedent provided by the swap agreement is a possible price reduction, the cedent may be unwilling to transfer profits. On the contrary, the arrangement may be interesting for the reinsurer, depending on the hedging tools available (so that it could even be the only arrangement available on the reinsurance market). See Sect. 4.6.7 in this regard.

The design of the reinsurance-swap can be generalized by assigning two barriers Λ_t', Λ_t'' (with $\Lambda_t' \leq B_t^* \leq \Lambda_t''$), such that:

$$B_t^{[\text{swap}]} = \begin{cases} B_t - \Lambda_t' & \text{if } B_t \leq \Lambda_t' \\ 0 & \text{if } \Lambda_t' < B_t \leq \Lambda_t'' \\ B_t - \Lambda_t'' & \text{if } B_t > \Lambda_t'' \end{cases} \tag{4.33}$$

The net outflow (gross of the reinsurance premium) for the cedent is then:

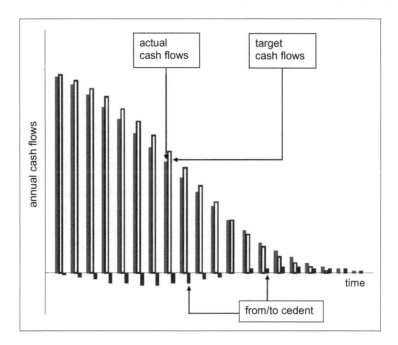

Fig. 4.48 Cash flows in a Longevity Swap arrangement

$$O F_t^{[\text{swap}]} = B_t - B_t^{[\text{swap}]} = \begin{cases} \Lambda_t' & \text{if } B_t \le \Lambda_t' \\ B_t & \text{if } \Lambda_t' < B_t \le \Lambda_t'' \\ \Lambda_t'' & \text{if } B_t > \Lambda_t'' \end{cases} \qquad (4.34)$$

We note that implementing the flow defined by Eq. (4.34) results in a transfer to
the reinsurer of both large losses as well as large profits. Therefore, both a floor and a
cap are applied to profits/losses of the cedent. Money transfers in a swap arrangement
are sketched in Fig. 4.49.

Remark The swap arrangements described above do not aim at a reduction of the
insurer's cash flows. Actually, the target is to replace the original cash flows with
"smoother" cash flows: the replacement implies in some cases a reduction and in
other cases an increase with respect to the original cash flows.

4.6.7   Longevity Bonds

Longevity bonds belong to the class of ILS (see Sect. 4.5.7). A reference population
is chosen, whose mortality rates are observed along the lifetime of the bond. The
population typically consists of a given cohort. A mortality or a survivor index

Fig. 4.49 Cash flows in a longevity swap

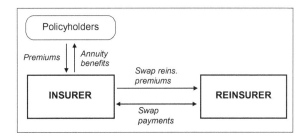

is defined, whose performance is assessed according to mortality in the reference population. We note that, to avoid lack of confidence in the way the pay-off of the bond is determined, mortality data should be collected and worked out by independent analysts; so, typically general population mortality data are referred to instead of insurance market data.

The coupon of the bond is contingent on the above index. In some cases, the principal of the bond may be reduced according to the behavior of the index.

Remark 4.1 It is worth noting the difference between longevity bonds and (fixed-income) long-term bonds. While the former are ILS whose performance is linked to some longevity index, the latter are traditional bonds with, say, a 20–25 years maturity, and (usually) a fixed annual interest (or possibly an annual interest linked to some economic or financial index, as for example an inflation index). Although not tailored to the specific needs arising from the longevity risk, long-term bonds can help in meeting obligations related to a life annuity portfolio. Actually, one of the most important problems in managing life annuities (with a guaranteed benefit) consists in mitigating the investment risk through the availability of fixed-income long-term assets, matching the long-term liabilities. Clearly, this problem becomes more dramatic as the expected duration of the life annuities increases thanks to mortality improvements.

Depending on its design, the longevity bond may offer hedging opportunities to a life annuity provider through either a long or a short position. As noted in Sect. 4.5.7, in the first case, the pay-off of the bond increases with decreasing mortality; vice versa in the second case. Given the long term maturity, in any case it is reasonable that the link is realized through the coupon, hence providing hedging on a yearly basis. In the following, we therefore assume that the principal does not depend on the experienced mortality. The reference population should be a given cohort, possibly close to retirement, i.e. with age 60–65 at bond issue.

Let L_t denote the number of individuals in the cohort after t years from issue, $t = 0, 1, \ldots$; viz. $L_0 = \ell_0$ is a known number. A maturity T may be chosen for the bond, with T high (for example, $T \geq 90 -$ initial age). In the following examples, some possible designs for the coupons are examined.

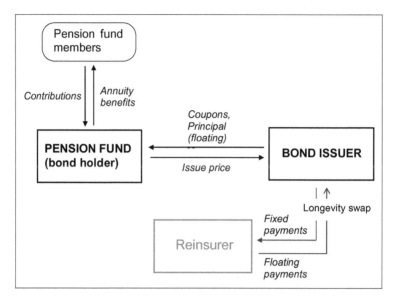

Fig. 4.50 Flows in a risk transfer via longevity bond—type 1 and longevity swap

Example 4.8 (*Longevity bond—Type 1*) The easiest way to link the coupon at time
t, C_t, to the longevity experience up to time t in the reference population is to let it
be proportionate to the observed survival rate. We then define:

$$C_t = C \frac{L_t}{\ell_0} \tag{4.35}$$

where C is a given amount. Hence, in case of unanticipated longevity the coupon
decreases slower than expected; so, a long position should be taken by an insurer
or pension fund dealing with life annuities. A simplified structure is sketched in
Fig. 4.50, where a reinsurer is also involved to hedge, via a longevity swap, the
longevity risk taken by the bond issuer. A similar bond was proposed by EIB/BNP
Paribas, however not traded on the market.   ∎

Example 4.9 (*Longevity bond—Type 2a*) Similarly to the mortality bond (see
Examples 4.6 and 4.7), two thresholds may be assigned, expressing survival levels. If the number of survivors in the cohort overperforms such thresholds, then the
amount of the coupon is reduced, possibly to 0. The following definition can be
adopted:

$$C_t = C \times \begin{cases} \dfrac{\ell_t'' - \ell_t'}{\ell_0} & \text{if } L_t \leq \ell_t' \\[2mm] \dfrac{\ell_t'' - L_t}{\ell_0} & \text{if } \ell_t' < L_t \leq \ell_t'' \\[2mm] 0 & \text{if } L_t > \ell_t'' \end{cases} \tag{4.36}$$

where C is a given amount, and ℓ_t', ℓ_t'' are the two thresholds, expressing given numbers of survivors; for example:

$$\ell_t' = \lambda_t' \, \mathbb{E}[L_t|S] \tag{4.37a}$$
$$\ell_t'' = \lambda_t'' \, \mathbb{E}[L_t|S] \tag{4.37b}$$

where $1 \leq \lambda_t' < \lambda_t''$ and S is a given mortality assumption on the reference cohort. Note that in this case the lower is mortality (i.e. the higher is L_t), the lower is the amount of the coupon. A short position should then be taken to hedge life annuities outflows. The structure of flows is like the one sketched in Fig. 4.38. ∎

Example 4.10 (*Longevity bond—Type 2b*) The coupon can be set proportionate to the number of deaths observed in the reference cohort from the bond issue. For example:

$$C_t = C \, \frac{\ell_0 - L_t}{\ell_0} \tag{4.38}$$

where $\ell_0 - L_t$ is the observed number of deaths up to time t. Clearly, also in this case a short position should be taken to hedge longevity risk. Again, the structure of flows is like the one shown in Fig. 4.38. ∎

Remark 4.2 We address some market issues, stressing in particular some significant difficulties.

1. A first issue concerns who can be interested in investing in bonds (types 2a and 2b) offering hedging opportunities to insurers/reinsurers. In general terms, one could argue that such securities may offer diversification opportunities, in particular thanks to their low correlation with standard risk causes in financial markets. Further, they may give long-term investment opportunities, otherwise scarcely when not at all available.
2. As regards bonds of type 1, from the point of view of the issuer, the possibility of building a longevity bond depends on the availability of financial securities with a proper maturity to match the payments promised under the longevity bond.
3. A further issue, already addressed, concerns the choice of mortality data. To encourage confidence on the linking mechanisms, reference to insurance market data should be avoided. Data recorded and analyzed by an independent institution should then be adopted. This raises a basis risk for hedgers, as the mortality in the reference population might differ from the mortality in life annuity portfolios.
4. Another aspect concerns difficulties in assessing and pricing the longevity risk transferred to the capital market. First of all, a generally accepted model for stochastic mortality is not yet available. Second, a market is not yet developed, nor similar risks are traded on the market itself. So, even in case of common agreement on a pricing model, data to estimate the relevant parameters are not yet available. Various approaches have been tested in the technical literature, research is in progress, but open issues still require careful investigation.

4.6.8 Hedging via Longevity Bonds: Examples

We present some examples of longevity hedging via longevity bonds. In particular, we consider the following cases:

- an insurer (or a pension fund) hedges the longevity risk inherent in the payment of the annuity benefits B_t, by purchasing bonds of type 1, hence taking a long position (Example 4.11).
- a reinsurer wrote a reinsurance treaty with an insurer, and hedges the longevity risk inherent in the payment of $B_t^{[SL]}$ or $B_t^{[swap]}$ (as defined in Sects. 4.6.5 and 4.6.6) by issuing and selling longevity bonds of type 2a or 2b, thus taking a short position (Examples 4.12 and 4.13).

We note that issuing longevity bonds may be difficult for an insurer or other annuity provider (e.g., a pension fund), because of the complexity of the deal. What is reasonable is that some form of reinsurance is underwritten by annuity providers; the reinsurer, who makes business on a larger scale than an insurer, hedges its position through longevity bonds (typically issued by a Special Purpose Vehicle).

In the following examples, we assume that all the annuitants receive the same annual benefit b. We denote by N_t the random number of annuitants alive at time t; n_0 denotes the given initial size of the annuitants cohort. We then have $B_t = b N_t$, $t = 1, 2, \ldots$.

Example 4.11 An insurer (or a pension fund) buys k units of a longevity bond type 1, and then will cash at time t the amount $k C_t = k C \frac{L_t}{\ell_0}$ (see Eq. (4.35)). The insurer's net outflow, for $t = 1, 2, \ldots$, is then:

$$OF_t^{[LB]} = B_t - k C \frac{L_t}{\ell_0} \tag{4.39}$$

If, for $t = 1, 2, \ldots$, we have:

$$\frac{N_t}{n_0} = \frac{L_t}{\ell_0} \tag{4.40}$$

then:

$$OF_t^{[LB]} = \frac{L_t}{\ell_0} (b n_0 - k C) \tag{4.41}$$

The outflow is still random because depending on L_t. Nonetheless, if

$$k = \frac{b n_0}{C} \tag{4.42}$$

then:

$$OF_t^{[LB]} = 0; \quad t = 1, 2, \ldots \tag{4.43}$$

and hence a perfect hedging is achieved. However, it is important to point out the following problems.

- In practical terms, a perfect hedging is difficult to realize. Although one can rely on some positive correlation between the survival rate in the reference population, $\frac{L_t}{\ell_0}$, and that in the annuitants' cohort, $\frac{N_t}{n_0}$, it is unrealistic that they coincide in each year (condition (4.40)), due to the fact that usually the annuitants do not constitute the reference population. In particular, the year of birth of the reference cohort and of annuitants may differ. This mismatching leads to basis risk in the strategy for managing longevity risk.
- A second aspect concerns the lifetime of the bond. Typically the bond is not issued when life annuity payments start. If issued earlier, the problem would consist in the availability of the bond, in the required size, in the secondary market. If issued later than life annuities, the longevity risk of the insurer would be unhedged for some years. Moreover, a critical aspect is the lifetime of the bond: its maturity not necessarily coincides with the time of complete runoff of the annuitants cohort. ∎

Example 4.12 We assume that the insurer has written a Stop-loss reinsurance arrangement on the outflows, and the reinsurer hedges the related longevity risk by issuing a bond type 2a. The reinsurer's total outflow, $F_t^{[SL]}$, is then:

$$F_t^{[SL]} = B_t^{[SL]} + k\,C_t = \begin{cases} 0 & \text{if } b\,N_t \leq \Lambda_t' \\ b\,N_t - \Lambda_t' & \text{if } \Lambda_t' < b\,N_t \leq \Lambda_t'' \\ \Lambda_t'' - \Lambda_t' & \text{if } b\,N_t > \Lambda_t'' \end{cases}$$

$$+ k\,C \times \begin{cases} \frac{\ell_t'' - \ell_t'}{\ell_0} & \text{if } L_t \leq \ell_t' \\ \frac{\ell_t'' - L_t}{\ell_0} & \text{if } \ell_t' < L_t \leq \ell_t'' \\ 0 & \text{if } L_t > \ell_t'' \end{cases} \tag{4.44}$$

To achieve a perfect hedging, the thresholds in the reinsurance treaty must be stated consistently with the features of the longevity bonds:

$$\Lambda_t' = \frac{\ell_t'}{\ell_0}\,b\,n_0 \tag{4.45a}$$

$$\Lambda_t'' = \frac{\ell_t''}{\ell_0}\,b\,n_0 \tag{4.45b}$$

Then, we can rewrite the total outflow as follows:

$$F_t^{[SL]} = b\,n_0 \times \begin{cases} 0 & \text{if } \frac{N_t}{n_0} \leq \frac{\ell_t'}{\ell_0} \\ \frac{N_t}{n_0} - \frac{\ell_t'}{\ell_0} & \text{if } \frac{\ell_t'}{\ell_0} < \frac{N_t}{n_0} \leq \frac{\ell_t''}{\ell_0} \\ \frac{\ell_t'' - \ell_t'}{\ell_0} & \text{if } \frac{N_t}{n_0} > \frac{\ell_t'}{\ell_0} \end{cases}$$

$$+ k\,C \times \begin{cases} \frac{\ell_t'' - \ell_t'}{\ell_0} & \text{if } \frac{L_t}{\ell_0} \leq \frac{\ell_t'}{\ell_0} \\ \frac{\ell_t'' - L_t}{\ell_0} & \text{if } \frac{\ell_t'}{\ell_0} < \frac{L_t}{\ell_0} \leq \frac{\ell_t''}{\ell_0} \\ 0 & \text{if } \frac{L_t}{\ell_0} > \frac{\ell_t''}{\ell_0} \end{cases} \tag{4.46}$$

If $\dfrac{N_t}{n_0} = \dfrac{L_t}{\ell_0}$, then Eq. (4.46) reduces to:

$$F_t^{[SL]} = \begin{cases} k\,C\,\dfrac{\ell_t''-\ell_t'}{\ell_0} & \text{if } \dfrac{L_t}{\ell_0} \le \dfrac{\ell_t'}{\ell_0} \\[2mm] b\,n_0\,\dfrac{L_t-\ell_t'}{\ell_0} + k\,C\,\dfrac{\ell_t''-L_t}{\ell_0} & \text{if } \dfrac{\ell_t'}{\ell_0} < \dfrac{L_t}{\ell_0} \le \dfrac{\ell_t''}{\ell_0} \\[2mm] b\,n_0\,\dfrac{\ell_t''-\ell_t'}{\ell_0} & \text{if } \dfrac{L_t}{\ell_0} > \dfrac{\ell_t''}{\ell_0} \end{cases} \qquad (4.47)$$

Further, if $k = \dfrac{b\,n_0}{C}$, then:

$$F_t^{[SL]} = \begin{cases} b\,n_0\,\dfrac{\ell_t''-\ell_t'}{\ell_0} & \text{if } \dfrac{L_t}{\ell_0} \le \dfrac{\ell_t'}{\ell_0} \\[2mm] b\,n_0\,\dfrac{\ell_t''-\ell_t'}{\ell_0} & \text{if } \dfrac{\ell_t'}{\ell_0} < \dfrac{L_t}{\ell_0} \le \dfrac{\ell_t''}{\ell_0} \\[2mm] b\,n_0\,\dfrac{\ell_t''-\ell_t'}{\ell_0} & \text{if } \dfrac{L_t}{\ell_0} > \dfrac{\ell_t'}{\ell_0} \end{cases}$$

$$= b\,n_0\,\dfrac{\ell_t''-\ell_t'}{\ell_0} \qquad (4.48)$$

Hence, $F_t^{[SL]}$ is a flow certain. We note that the assumptions underpinning the perfect hedging are the same adopted for the longevity bond type 1 (see Example 4.11); the same remarks then apply.   ∎

Example 4.13 We assume that the insurer has written a longevity swap arrangement on the outflows, and the reinsurer hedges the related longevity risk by issuing a bond type 2b. The reinsurer's total outflow, $F_t^{[swap]}$, is then:

$$F_t^{[swap]} = B_t - B_t^* + k\,C_t \qquad (4.49)$$

Let

$$n_t^* = \mathbb{E}[N_t \mid S] \qquad (4.50)$$

where S denotes, as usual, the mortality scenario. Assume that the target outflow, B_t^*, for the swap arrangement (see Eq. (4.30)) can be expressed as follows:

$$B_t^* = \mathbb{E}[B_t \mid S] = b\,\mathbb{E}[N_t \mid S] = b\,n_t^* \qquad (4.51)$$

Hence, the net flow for the reinsurer can be rewritten as follows:

$$F_t^{[swap]} = b\,n_0\,\dfrac{N_t}{n_0} - b\,n_t^* + k\,C\,\dfrac{\ell_0 - L_t}{\ell_0}$$

$$= k\,C - b\,n_t^* + b\,n_0\,\dfrac{N_t}{n_0} - k\,C\,\dfrac{L_t}{\ell_0} \qquad (4.52)$$

If $\dfrac{N_t}{n_0} = \dfrac{L_t}{\ell_0}$ and $k = \dfrac{b\,n_0}{C}$, then Eq. (4.52) reduces to:

$$F_t^{[\text{swap}]} = k\,C - b\,n_t^* = b\,(n_0 - n_t^*) \tag{4.53}$$

which again expresses a situation certain. Remarks on the possibility of realizing a perfect hedging are as in the previous cases. ∎

4.6.9 Natural Hedging

In the context of life insurance, natural hedging refers to a diversification strategy realized by combining "opposite" benefits with respect to the duration of life (path $(9) \rightarrow (b)$ in Fig. 4.44). The underlying idea is that if mortality rates decrease then life annuity payments increase while death benefit payments decrease (and vice versa if mortality rates increase). Hence, the longevity risk inherent in a life annuity business could be offset, at least partially, by taking a position also on some insurance products providing benefits in case of death. We discuss two situations, one concerning hedging across time and one across lines of business (LOBs).

We first consider *hedging across time*. We assume that the life annuity product provides capital protection (see Sect. 5.5). Hence, in the case of early death of the annuitant, the difference (if positive) between the single premium and the cumulated benefits paid to the annuitant will be paid to the beneficiaries. Intuitively, when dealing with both life annuity benefits and a death benefit the insurer gains a risk reduction, given that the longer are the annuity payment periods, the lower are the amounts of death benefit.

A diversification could be pursued via *hedging across LOBs*, that is, by properly mixing positions in life assurances and life annuities. The effectiveness of course depends on the mortality trend over the whole range of ages involved. Actually, life assurances usually concern a different range of ages than life annuities (in particular, if term assurances are only involved). If mortality trends emerge differently within life assurance and life annuity blocks of business, then the effectiveness may be rather poor. Moreover, to achieve a significant risk reduction, the magnitude of the costs of life assurances should be similar to those of life annuities. A proper diversification effect between life assurance and life annuities may be difficult to obtain by an insurer on its own, whereas a more effective offset can be achieved by a reinsurer.

4.6.10 Bulk Annuities: Buy-In, Buy-Out

Pension funds providing life annuity benefits are exposed to several risks: in particular, the longevity risk and the financial risk. A longevity swap agreement (see Sect. 4.6.6) between a pension fund and an insurer (rather than between an insurer and a reinsurer) can be used to transfer (to an insurer) the longevity risk (and, in particular, its systematic component). Other risk transfers are outside the scope of the longevity swap.

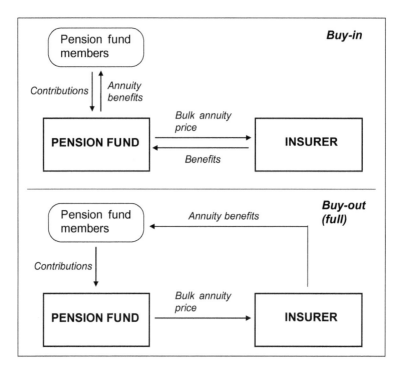

Fig. 4.51 Bulk annuities: buy-in and buy-out schemes

All the risks inherent in life annuity payout (and hence longevity risk as well as financial risk, liquidity risk, etc.) can be transferred to the insurance market via *bulk annuities*, hence de-risking the pension fund.

Bulk annuities are long-term insurance policies provided by life insurers, bespoke for a given pension plan. Bulk annuities can be split into two categories, according to the way they transfer risks from the pension fund to the insurer.

- In a *buy-in* arrangement, the pension fund purchases a policy from an insurer to match the annuity benefits paid by the fund itself. The policy can cover all or just part of the pensioner population, but can also include future pensioners; the policy is held as an asset of the pension fund. The pension plan remains in place.
- A *buy-out* aim at extinguishing pension fund liabilities, with the insurer issuing individual annuity policies to members of the pension plan. Usually this arrangement covers all the liabilities of the pension plan (full buy-out) and, in this case, the wind-up of the pension plan follows.

The flows in the two bulk annuity arrangements are sketched in Fig. 4.51.

4.7 References and Suggestions for Further Reading

A number of papers and technical reports deal with ERM for insurance business and pension funds. The reader is referred, for example, to IAA (2011, 2016, 2010a), Koller (2011), and Calandro (2006).

Risk identification and, in particular, risk causes, components and factors are described in IAA (2004). The distinctive features of risk and uncertainty constitute the topic addressed by Gutterman (2017) in the IAA Risk Book.

An introduction to stochastic approaches for life insurance and life annuity portfolios is provided by Pitacco (2007), while special attention is placed on managing longevity risk in life annuity business by Pitacco et al. (2009). A stochastic model for an endowment insurance portfolio is proposed by Olivieri and Pitacco (2008), in the framework of life insurance evaluations.

Stress testing and scenario analysis are focused by IAA (2013), and in the IAA Risk Book by Müller and Sandberg (2020). An example of sensitivity analysis is provided by Pitacco (2016b), where packaging lifetime and long-term care benefits is addressed (in this regard, see also Chap. 8).

The papers by Olivieri and Pitacco (2002, 2012) focus on stochastic mortality and proposes a mortality monitoring procedure based on Bayesian inference, while Olivieri and Pitacco (2009b) deals with mortality monitoring in the framework of capital allocation modeling.

Solvency issues and related capital requirements are addressed by many papers and technical reports. The reader is in particular referred to IAA (2009, 2011), and, regarding the use of internal models, IAA (2010b). Capital requirements for life annuity portfolios according to internal models and short-cut formulae are compared by Olivieri and Pitacco (2009a).

The reader interested in reinsurance arrangements for life insurance and life annuity business is referred to Tiller and Fagerberg Tiller (2015). Non-proportional reinsurance (for life and non-life business) is addressed by Eves et al. (2015) in the IAA Risk Book. The book by Booth et al. (2005) also focuses on the risks inherent in a life insurance portfolio and the related hedging actions, reinsurance included. An extensive and detailed presentation of Alternative Risk Transfers and Insurance-linked securities is provided, in particular, by Blake et al. (2006). Longevity risk transfers, including longevity swaps and bulk annuities, and related market issues are addressed in Blake et al. (2019). The paper by Zhu and Bauer (2014) focuses on natural hedging.

Within the broader area of insurances of the person, the following papers deal with ERM in health insurance: Orros and Howell (2008), Orros and Smith (2010), and Rudolph (2009).

Chapter 5
Product Design: Guarantees and Options

5.1 Outlook

The following Chaps. 6–8 focus on two phases of the RM process, that is, on risk assessment and impact assessment. Numerical results basically consist of the output of different modeling approaches and assumptions on the scenarios. These results may be delivered by an actuary to the risk managers who will use them for an appropriate choice of RM actions.

Modeling risk assessment and impact assessment is frequently considered the only (or, at least, the most important) task of the actuarial job in insurance companies. However, as noted in Sect. 3.2.2, the ERM logic calls for appropriate interactions among the various functions and processes in any organization, the product development department in particular (see Fig. 3.3).

As far as the development of insurance products is concerned, especially (but not only) in the area of insurances of the person, also the product design phase must rely on careful technical assessments (besides obvious market consideration), hence requiring a substantial actuarial intervention, notably in terms of actuarial "mentality", that is, ability to capture ab initio all the risk features of the product.

Issues related to product development and product design will be presented in two steps.

- In the present chapter we focus on product features which should be carefully considered when designing and launching new products, as well as when assessing the risk profile of existing products. In the former case, a specific phase, requiring a substantial actuarial involvement, must be introduced into the RM process.
- In Chap. 9, the new product development process will specifically be addressed referring to the design and the launch of life annuity products.

E. Pitacco, *ERM and QRM in Life Insurance*, Springer Actuarial,
https://doi.org/10.1007/978-3-030-49852-8_5

5.2   The Product Design as a RM Action

The RM process sketched in Fig. 3.1, in the framework of life insurance and life annuity business, can refer to an insurance company or a portfolio (or line of business). When referred to a company, the RM process is at the heart of ORSA, as noted in Sect. 3.3. When referred to a portfolio, it is assumed that the portfolio itself consists of existing products, that is, products whose features have previously been defined. However, the same logical structure can be adopted when referring to a line of business consisting of new products whose structure definition is in progress. New ingredients then enter into the process, and the scheme must be extended accordingly.

The *new product development* (*NPD*) process and then the launch of a new line of products call for a "preliminary" implementation of the RM process, which starts from the *product design*. Then, the usual RM phases follow, however extended to include, among the RM actions, the possible product re-design, as sketched in Fig. 5.1.

Many modern insurance products are designed as packages, whose items may be either included or not in the product actually purchased by the client. For example: the endowment insurance which can include various rider benefits and options, the Universal Life insurance, the Variable Annuities, the presence of possible long-term care benefits in pension products.

The benefits provided by these products imply a wide range of guarantees and hence risks borne by the insurance company (or the pension fund). Guarantees can be classified as follows:

- *implicit guarantees* (or *embedded guarantees*), included in the basic product as a consequence of fixed policy conditions;

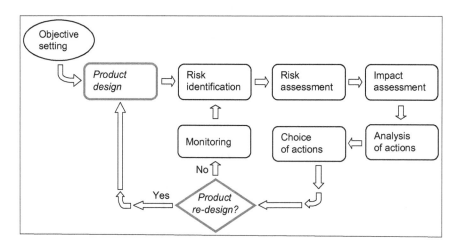

Fig. 5.1   The product design in the life insurance RM process

- *explicit guarantees*, as the result of specific options exercised by the policyholder, at the policy issue or throughout the policy duration.

Risks originated by guarantees clearly emerge in recent scenarios, in particular because of volatility in the financial markets and trends in mortality/longevity. In the risk identification phase all relevant risks are listed, and then modeled in the risk assessment phase. Impacts on results of interest are conveniently modeled in the impact assessment phase.

Results achieved in the impact assessment phase must drive the choice of appropriate RM actions. Among the various actions, *pricing* of the product deserves special attention. While a "tentative" pricing can provide a provisional input to the impact assessment phase (e.g., to check the profitability of the product, in line with the logic of profit testing), a sound pricing should account for randomness in the results of interest.

Appropriate modeling tools should be used for pricing (and reserving). Hence, a shift from expected present values to more modern and complex approaches is, at least in principle, needed. However, the implementation of complex mathematical methods often constitutes, on the one hand, an obstacle on the way towards sound pricing principles. On the other hand, facing the risks by charging very high premiums might reduce the insurer's market share. Then, alternative solutions can be suggested by appropriate product designs which aim at sharing risks between the insurer and the policyholders, thus weakening (implicit) guarantees. An interesting example is provided by profit participation mechanisms regarding financial market risks (see Sect. 6.2).

Results obtained in the above phases might suggest the product re-design (and pricing), before the launch of the product itself (see Fig. 5.1). Re-design of the product might be motivated and suggested, for example, by the following reasons.

- The premium of the product (as results from the pricing procedure) is too high if compared to premiums charged by competitors in the insurance market; hence, a poor market share might follow.
- According to solvency regime, the product is too "capital absorbing". A "light capital" product can be obtained by removing some guarantees.

Guarantees (to be meant as implicit guarantees) and options (whose exercise implies further guarantees), and related risks, in some life insurance and life annuity products are listed and commented in the following sections.

5.3 Term Insurance

We first describe the main features of term insurance products. We then focus on some underwriting issues, and mortality risk transfers.

5.3.1   Product Main Features

The most important guarantees provided by a term insurance, as well as some options which can be included in this product, are illustrated in Fig. 5.2. The *mortality guarantee* implies that, whatever the number of deaths in the portfolio, the insurer has to pay the death benefit amount as stated in the policy. According to the *interest guarantee*, the policy reserve must be annually credited with an amount calculated with the specified interest rate, whatever the investment yield obtained by the insurer. It is worth noting that, because of the relatively small amount of the policy reserve in the term insurance, this guarantee does not have a dramatic impact on portfolio results even in the case of very poor investment yield.

Various options can be included in the term insurance policy. For example, policyholders can choose among several *settlement options* regarding the payment of the death benefit. In particular:

- usually the benefit is paid to the beneficiary as a lump sum;
- as an alternative, the benefit can be paid during a fixed period as a sequence of installments;
- another alternative consists in paying the benefit as a life annuity to the beneficiary, as long as he/she is alive; it is worth stressing that, in this case, a longevity risk is taken by the insurer.

The second and third alternative are usually denoted as *structured settlements*.

According to *guaranteed insurability* (or *benefit increase option*), the policyholder may apply for an increase in the sum assured in face of some specific events, typically concerning his/her household, such as the birth of a child, without being adopted a revised mortality basis. The risk implied by possible adverse selection is hence taken by the insurer.

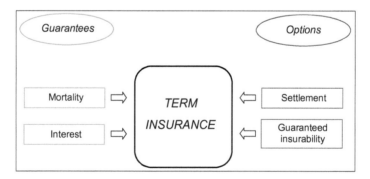

Fig. 5.2 Guarantees and options in a term insurance policy

5.3.2 Underwriting Issues

Term insurance business involves *underwriting requirements* which aim at assessing the risk exposure of the applicant. When a death benefit is assured, health conditions, occupation and smoking status can be taken as rating factors. These lead to a classification into *standard risks* and *substandard risks*. For the latter (also referred to as *impaired lives*), a higher premium level is adopted in order to avoid (or, at least, to reduce) adverse selection, given that they bear a higher probability to become eligible for the benefit. In some markets, standard risks are further split into *regular* and *preferred risks*, the latter having a better profile than the former (for example, because they never smoked); as such, they are charged a reduced premium rate.

Diverse underwriting schemes are adopted around the world. To summarize, the following classification can be adopted.

- *Fully underwritten business*. Underwriting requirements consist of a medical (or paramedical) exam and a set of medical questions.
- *Simplified issue business*. The underwriting is only based on a set of medical questions.
- *Guaranteed issue business*. Neither medical (or paramedical) exam nor medical questions are involved; can be applied for small amounts assured.

5.3.3 Mortality Risk in Term Insurance Portfolios

The presence of mortality risk at a portfolio level must be hedged via appropriate RM actions. Besides capital allocation, appropriate risk transfers can be implemented via reinsurance arrangements, at policy or at portfolio level (see Sects. 4.5.3–4.5.5).

In particular:

- reinsurance on risk-premium basis and individual surplus on original terms can be used to mitigate mortality risk by leveling the sums assured;
- Cat-XL reinsurance can help in lowering the impact of extreme events.

5.4 Endowment Insurance

After describing the main features of the endowment insurance product, we focus on interest guarantees related to participation mechanisms.

5.4.1 Product Main Features

Some guarantees and options provided by an endowment insurance policy are similar to the corresponding guarantees and options in the term insurance (in particular, the *mortality guarantee* and the *settlement options*). Conversely, the risk implied by the *interest guarantee* is much higher than in the term insurance, because of the important amount progressively accumulated through the reserving process.

Several options can be included in an endowment insurance policy. The following ones are of particular interest (see Fig. 5.3). If the *surrender option* is exercised, the contract terminates and the surrender value (that is, the amount of the policy reserve minus the surrender fee) is paid to the policyholder. Several risks are implied by this option; for example, the market risk (when the insurer is forced to sell bonds with an interest rate lower than the current rate), the liquidity risk, etc.

In participating products, various *dividend options* can be available, which allow the policyholder to participate in insurer's profits (arising from investment return, mortality, expenses). In particular:

1. dividends can be paid in cash, usually via reduction of future premiums;
2. as an alternative, frequently adopted in many European policies, dividends can be used to finance increments in the sum insured (either in the case of survival at maturity, or in the case of death, or both);
3. another alternative consists in a financial accumulation of the dividends, with a guaranteed interest rate.

Alternatives 3 and, possibly, 2 (according to the mechanism adopted for increasing the sum at maturity) imply a financial risk borne by the insurer; we will analyze some relevant aspects in Sect. 6.2.

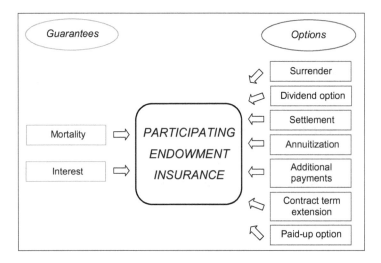

Fig. 5.3 Guarantees and options in a participating endowment policy

If the *annuitization option* is exercised, the survival benefit is paid as a life annuity, i.e., as long as the beneficiary is alive, instead of being paid as a lump sum at maturity. A crucial problem is related to the time at which the annuitization rate is stated; this time can vary from the date of the endowment policy issue to policy maturity: the sooner this rate is fixed, the higher is the aggregate longevity risk, due to the uncertainty in future mortality trend, taken by the insurer. Anyway, whatever the time at which the annuitization rate is stated, if the annuitization option is exercised, various risks are taken by the insurer, and in particular:

- the adverse selection risk, caused by the likely good health conditions of the beneficiary who annuitizes, and hence by a presumably long expected lifetime;
- the aggregate longevity risk;
- the financial risk, originated by the minimum interest guarantee usually provided by the life annuity.

By exercising the *additional payments option*, the policyholder can increase the sum assured. As regards the death benefit, this option implies the *guaranteed insurability* (see Sect. 5.3).

Thanks to the *contract term extension*, the policyholder can take advantage from the guaranteed interest rate; thus, the value of this option depends on the current interest rate. If exercised, the financial risk taken by the insurer extends beyond the original policy maturity.

The *paid-up option* is exercised when the policyholder stops the premium payment without terminating the insurance contract. Thus, the contract remains in force with properly reduced benefits.

5.4.2 Participation Mechanisms

Financial risk is, because of the interest guarantee, an important item of the risk profile of an endowment insurance portfolio. Appropriate investment strategies must be implemented to provide the insurer with a reasonable yield, while keeping the volatility of results at a low level.

Guaranteed interest rates are currently very low (when not equal to zero), so that participation mechanisms work to provide the insured with a share of the possible extra-yield. A diversity of mechanisms, aiming to raise the insured benefits via increase in the policy reserve, are available. A significant diversity of portfolio risk profiles follow.

Hence, the choice of a specific participation mechanism deserves great attention. This topic will be addressed in Sect. 6.2.

5.5 Immediate Life Annuities

While the (standard) immediate life annuity is the product we are going to focus, basic features of all life annuity products are first recalled and discussed. Some issues concerning longevity risk are then addressed. Product designs aiming at a raise in the potential number of annuitants are finally presented.

5.5.1 Product Main Features

When planning the post-retirement income, some basic features of the life annuity product should carefully be considered by any individual. In particular, we recall the following aspects.

1. The life annuity product relies on the mutuality mechanism. This means that:
   a. the amounts released by the deceased annuitants are shared, as mortality credits, among the annuitants who are still alive;
   b. on the annuitant's death, his/her estate is not credited with any amount, and hence no bequest is available.

2. A life annuity provides the annuitant with an "inflexible" post-retirement income, in the sense that the annual amounts must be in line with the benefit profile, as stated by the policy conditions.
3. Purchasing a life annuity is an irreversible decision: surrendering is generally not allowed to annuitants (clearly, to avoid adverse selection effects; see Sect. 4.2.7). Hence, the life annuity constitutes an "illiquid" asset in the retiree's estate.

Features 1(b), 2 and 3 can be perceived as disadvantages, and can hence weaken the propensity to immediately annuitize the whole amount available at retirement. These disadvantages can be mitigated, at least to some extent, either by purchasing life annuity products in which other benefits are packaged, or adopting specific annuitization strategies.

All the above features should be carefully considered by the insurer while planning the launch of a life annuity product.

The range of guarantees and options provided by life annuities and the relevant features are strictly related to the type of life annuity product. For example, in a deferred life annuity both the accumulation and the decumulation (or payout) phases are involved, so that some guarantees (e.g., the interest rate guarantee) can extend over a period of several decades. Moreover, the amount of longevity risk borne by the insurer (or, in general, by the annuity provider) depends on the time at which the annuitization rate is stated. In a traditional deferred life annuity, the annuitization rate and hence the annuity benefit are stated at the policy inception, namely at the beginning of the accumulation phase; this implies a substantial amount of aggregate longevity risk taken by the insurer. Conversely, if the annuitization rate is stated at

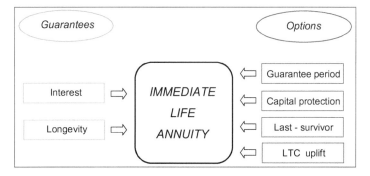

Fig. 5.4 Guarantees and options in an immediate life annuity

the end of the accumulation period, a smaller amount of longevity risk is borne by the insurer.

For brevity, we only focus on the decumulation phase, thus addressing immediate life annuities (see Fig. 5.4).

The *interest guarantee* has already been discussed in Sects. 5.3 and 5.4. Of course, in a life annuity the importance of this guarantee is a consequence of the average long duration of the annuity itself. Thanks to the *longevity guarantee*, the annuitant has the right to receive the stated annuity benefit as long as he/she is alive, and hence:

1. whatever his/her lifetime;
2. whatever the lifetimes of the annuitants in the annuity portfolio (or pension fund).

Because of feature 1, the annuity provider takes the individual longevity risk, originated by random fluctuations of the individual lifetimes around the relevant expected values. Feature 2 also implies the aggregate longevity risk: if the average lifetime in the portfolio is higher than expected, the annuity provider suffers a loss, because of systematic deviations of the lifetimes from the relevant expected values.

Various options can be added to the life annuity product. These options can be exercised before the start of the payout period, that is, at policy issue or, in the case of deferred annuities, before the end of the deferment period (usually with some constraints, e.g., 6 months before the end of this period, to reduce possible adverse selection). By exercising these options, other benefits are added as riders to the basic life annuity product, and hence other guarantees provided.

A life annuity with *a guarantee period* (5 or 10 years, say), also named *period-certain life annuity*, pays the benefit over the guarantee period regardless of whether the annuitant is alive or not.

By exercising the *capital protection* (or *money-back*) option, a death benefit is added to the life annuity product, then usually called *value-protected life annuity*. In the case of early death of the annuitant, a value-protected annuity will pay to the annuitant's estate the difference (if positive) between the single premium and the cumulated benefits paid to the annuitant. Usually, capital protection expires at some given age (75, say), after which nothing is paid even if the above difference is positive.

A *last-survivor annuity* is an annuity payable as long as at least one of two individuals (the annuitants), say (x) and (y), is alive. It can be stated that the annuity continues with the same annual benefit, say b, until the death of the last survivor. A modified form provides that the amount, initially set to b, will be reduced following the first death: to b' if individual (y) dies first, and to b'' if individual (x) dies first, usually with $b' < b$, $b'' < b$. Conversely, in many pension plans the last-survivor annuity provides that the annual benefit is reduced only if the retiree, say individual (x), dies first. Formally, $b' = b$ and $b'' < b$. Whatever the arrangement, the expected duration of a last-survivor annuity is longer than that of an ordinary life annuity (that is, with just one annuitant). A higher amount of longevity risk (both individual and aggregate) is then borne by the annuity provider.

By exercising the *LTC (Long-Term Care) uplift* option, a health-related benefit is added to the basic life annuity. The resulting product, which is often called *enhanced pension*, is a combination of a standard life annuity paid while the policyholder is healthy, and an uplifted income paid while the policyholder is claiming for the LTC benefit. See also Sect. 5.6.2.

5.5.2 Hedging the Longevity Risk

Exposure to longevity risk and, in particular, to aggregate longevity risk is, by far, the most important item in the risk profile of a life annuity portfolio. Besides capital allocation (in particular required by solvency regimes) and possible transfer via traditional reinsurance and ARTs (see Sect. 4.6), specific product designs, sharing the common target of weakening the longevity guarantee, can significantly help in mitigating the longevity risk (and its aggregate component). We focus on the two following solutions.

A *temporary life annuity* provides the annuitant with a sequence of periodic benefits provided he/she is alive but up to a given age, 85 or 90 say, stated in the policy conditions (and hence at most for a given number of years). This way, the tail of the lifetime distribution is not involved and the longevity risk borne by the annuity provider consequently reduced. However, we note that this type of product, in spite of an obvious reduction in the premium, may be less attractive from the perspective of the client because of the "uncovered" age interval.

According to alternative product designs, part of the longevity risk can be transferred to the annuitants. This implies the definition of a *longevity-linked life annuity*. A longevity-linked life annuity involves a benefit adjustment process, according to which the annuity provider is entitled to reduce the benefit to all the annuitants in the event of an unanticipated increase in longevity. However, a floor amount should reasonably be stated to keep, at least to some extent, the guarantee characteristics which should feature all life annuity products.

Basic problems in defining the longevity-driven adjustment process are:

1. the choice of the age-pattern of mortality referred to;
2. the choice of the link between annual benefits and mortality.

These choices should reasonably aim at sharing the aggregate longevity risk (that is, the systematic component of the longevity risk), while leaving the volatility (the random fluctuation component) with the annuity provider, as the latter can be diversified by risk pooling, viz inside the traditional insurance—reinsurance process.

5.5.3 Raising the Number of Annuitants

As (standard) life annuities are attractive mainly for healthy people, in order to expand their business, in recent years some insurers have started offering better annuity rates to people whose health conditions are worse than those of likely annuity buyers. *Special-rate life annuity* products have then been designed.

Special-rate life annuities are also called *underwritten life annuities*, because of the ascertainment of higher mortality assumptions via the underwriting requirements. Details on special-rate life annuities will be provided in Sect. 9.3.1.

Premium reduction is generally a way to increase the attractiveness of any insurance product. Temporary life annuities (see Sect. 5.5.2) can be sold at a premium lower than that of a standard life annuity thanks to a restriction of the age interval covered, and hence insurer's lower exposure to the longevity risk (notably significant on the tail of the lifetime distribution).

Old-age life annuities (also known as *advanced life delayed annuities*[1]) can achieve a similar premium reduction effect thanks to a restriction in the opposite direction: an old-age life annuity pays a lifelong benefit but starting from a high age (75 or 80, say), and hence provides protection against the individual longevity risk. Of course, the annuity provider takes the exposure to the tail risk. A (temporary) drawdown process, from retirement time to the commencement time of the life annuity, will provide the retiree with a post-retirement income.

Remark We note that coverage restrictions are commonly adopted in many insurance lines of business, in particular to reduce premium amounts. So, in the context of life annuities the presence of a *deductible* can be recognized in old-age life annuities, whereas an *upper limit* is a feature of the temporary life annuity.

---

[1] The advanced life delayed annuity (briefly, ALDA) has been proposed by Milevsky (2005).

5.6   Long-Term Care Insurance Products

Long-term care insurance (LTCI) provides the insured with financial support, while he/she needs nursing and/or medical care because of chronic (or long-lasting) conditions or ailments.

Several types of benefits can be provided (in particular: fixed-amount annuities, care expense reimbursement). The benefit trigger is usually given either by claiming for nursing and/or medical assistance (together with a sanitary ascertainment), or by assessment of the individual disability, according to some predefined metrics, e.g., the Activities of Daily Living (ADL) scale, or the Instrumental Activities of Daily Living (IADL) scale. In different insurance markets, different criteria for the LTCI benefit eligibility are adopted (e.g., different ADL or IADL scales). This may complicate the problem of finding, for a given LTCI product, reliable statistical data concerning senescent disablement and mortality of disabled people.

5.6.1   LTCI Products: A Classification

LTCI products can be classified as follows:

- products which pay out benefits with a *predefined amount* (usually, a lifelong annuity benefit); in particular

  - a *fixed-amount* benefit;
  - a *degree-related* (or *graded*) benefit, i.e., a benefit whose amount is graded according to the degree of disability, that is, the severity of the disability itself (for example, assessed according to an ADL or IADL scale);

- products which provide reimbursement (usually partial) of nursery and medical expenses, i.e., *expense-related* benefits;
- *care service* benefits (for example, provided in the US by the Continuing Care Retirement Communities).

In what follows we only address LTCI products which pay out benefits with a predefined amount.

5.6.2   Fixed-Amount and Degree-Related Benefits

A list of LTCI products which pay out benefits with a predefined amount follows.

Immediate care plans, or *care annuities*, relate to individuals already affected by severe disability (that is, in "point of need"), and then consist of:

- the payment of a single premium;
- an immediate life annuity, whose annual benefit may be graded according to the disability severity.

Hence, care annuities are aimed at seriously impaired individuals, in particular persons who have already started to incur LTC costs. The premium calculation is based on assumptions of short life expectancy. However, the insurer may limit the individual longevity risk by offering a limited-term annuity, namely a temporary life annuity.

Pre-funded plans consist of:

- the accumulation phase, during which periodic premiums are paid (of course, while the insured is in the healthy state); as an alternative, a single premium can be paid;
- the payout period, during which LTC benefits (usually consisting of a life annuity) are paid in the case of LTC need.

Several products belong to the class of pre-funded plans.

A *stand-alone LTC cover* provides an annuity benefit, possibly graded according to an ADL or IADL score. This cover can be financed by a single premium, by temporary periodic premiums, or lifelong periodic premiums. Of course, premiums are waived in the case of an LTC claim. This insurance product only provides a "risk cover", as there is no certainty in future LTC need and the consequent payment of benefits.

A number of *combined products* (or *"combo" products*) have been designed, mainly aiming at reducing the relative weight of the risk component by introducing a "saving" component, or by adding LTC benefits to an insurance product with a significant saving component. Some examples follow.

LTC benefits can be added as a *rider to a whole-life assurance* policy (and hence can constitute an incentive against surrendering the policy). For example, a monthly benefit of, say, 2% of the sum assured is paid in the case of an LTC claim, for 50 months at most. The death benefit is consequently reduced, and disappears if all the 50 monthly benefits are paid. Thus, the (temporary) LTC annuity benefit is simply given by an *acceleration* of (part of) the death benefit. The LTC cover can be complemented by an additional deferred LTC annuity (financed by an appropriate premium increase) which will start immediately after the possible exhaustion of the sum assured (that is, if the LTC claim lasts for more than 50 months) and will terminate at the insured's death.

A *life insurance package* can include LTC benefits combined with lifetime-related benefits, i.e., benefits only depending on insured's survival and death. For example, the following benefits can be packaged:

1. a lifelong LTC annuity (from the LTC claim on);
2. a deferred life annuity (e.g., from age 80), while the insured is not in LTC disability state;
3. a lump sum benefit on death, which can alternatively be given by:

a. a fixed amount, stated in the policy;
b. the difference (if positive) between a stated amount and the amount paid as benefit 1 and/or benefit 2.

We note that item 2, i.e., the old-age deferred life annuity, provides protection against the individual longevity risk to people in good health conditions.

Life care pensions are life annuity products in which the LTC benefit is defined in terms of an uplift with respect to the basic pension b. In particular, the *enhanced pension* is a life care pension in which the uplift is financed by a reduction (with respect to the basic pension b) of the benefit paid while the policyholder is healthy. Thus, the reduced benefit b' is paid out as long as the retiree is healthy, while the uplifted benefit b'' will be paid in the case of an LTC claim (of course, $b' < b < b''$).

5.6.3 Guarantees, Options and Related Risks

As seen in the previous section, the class of LTCI products encompasses a broad range of covers. Guarantees and options provided, and hence risks taken by the insurer, vary moving from one product to another one.

For example, the scheme given by Fig. 5.5 can refer to a stand-alone product which, thanks to options, can be transformed into a combo product which includes lifetime-related benefits.

In particular, the stand-alone product provides the insured with the *disability guarantee*, which implies that, whatever the number of insureds entering into the LTC state, the insurer has to pay the benefits as stated in the policy conditions. Moreover, as the benefit is paid as an annuity, the *longevity guarantee* implies that the insurer has to pay lifelong benefits (provided that the insured remain in the LTC state). Disability risk and longevity risk are then taken by the insurer. The longevity risk of course refers to random lifetimes of insureds in the LTC state.

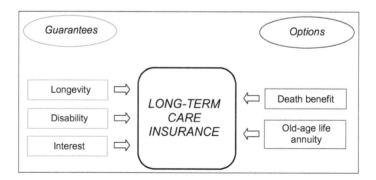

Fig. 5.5 Guarantees and options in a long-term care insurance policy

If the client decides to add a death benefit, then a mortality guarantee is also provided (as, for example, in a term insurance), and a mortality risk is consequently taken by the insurer.

An additional amount of longevity risk is added if the client decides to include an old-age life annuity into the benefit package.

It is worth noting that, regarding disability and longevity risk, the systematic component is particularly relevant in LTCI covers, given scarcity of data and hence uncertainty in the choice of appropriate technical bases. This topic will be addressed in Chap. 8.

5.7   From Liability-Driven to Asset-Driven Insurance Products

In fixed-benefit policies (for example: a term insurance or an endowment insurance without profit participation), the definition of liabilities (i.e., the benefits) comes before the selection of assets backing the policy reserve. These products can then be labeled as *liability-driven*.

As is well known, in unit-linked policies the asset perspective is prevailing. In particular, when no guarantee is provided by the policy, the insurer's liability is defined by the value of the units credited to the policy itself. Unit-linked policies are then considered to constitute an *asset-driven* business

The distinction reflects on the party bearing the financial risk, namely the insurer for liability-driven arrangements, the policyholder for asset-driven solutions.

Participating policies, as well as unit-linked policies with financial guarantees are somewhat at an intermediate step between a liability-driven and an asset-driven business. Basically, participating policies are liability-driven, as is suggested by the approach adopted for the calculation of premiums and reserves (see, for example, Sect. 6.1.1). However, the benefit amount and then the insurer's liability can be increased thanks to investment performance. Similarly, unit-linked policies with financial guarantees are basically asset-driven; however, since the guarantees transfer risk to the insurer, an *additional reserve* in respect of the unit fund may be necessary, which should be assessed consistently with the cost of the guarantee.

Figure 5.6 provides a graphical representation of the comments developed above. The large arrows show which is the starting point for the assessment of the value of assets and liabilities, or for their management: the liabilities for fixed-benefits and participating policies, the assets for unit-linked policies (with or without guarantees). In the case of participating policies, the small arrow expresses that the value of the liability must be updated according to the investment performance, while the small arrow in the case of unit-linked policies with guarantees recalls that the liability originated by the guarantee requires an appropriate hedging, and then an appropriate selection of assets.

Fig. 5.6 Liability-driven
versus asset-driven products

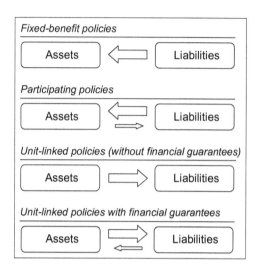

5.8 Guarantees in Unit-Linked Policies

In unit-linked policies (without guarantees) the financial risk is borne by the policy-holder. Partially, the risk can be transferred to the insurer, through the underwriting of appropriate guarantees.

Guarantees may relate to any of the benefits provided by the insurance cover:

- the *maturity guarantee* concerns the maturity benefit;
- the *death benefit guarantee* is given on the death benefit;
- the *surrender guarantee* concerns the surrender value.

In the following we refer to a unit-linked endowment insurance, with maturity at time m. We disregard the surrender guarantee, and we assume that the same type of guarantee is provided for the maturity and the death benefit. This way, we shorten a little bit the notation; anyhow, it is not difficult to address more general cases, in which the maturity and the death benefit guarantees are different.

The guarantee is defined specifying the minimum benefit amount. If, because of adverse financial trends, the policy fund at the time of payment is not high enough, the minimum amount will be paid.

Let B_t denote the benefit due at time t; if $t = 1, 2, \ldots, m - 1$ it is a death benefit, while if $t = m$ it is the benefit paid at maturity, either in case of death or survival (given that we are assuming that the same guarantee is provided for the maturity and the death benefit).

The guaranteed amount can be stated according to different targets. The simplest way is to set a fixed guaranteed amount G. Let F_t denote the policy fund at time t. The benefit at time t is then defined as follows:

$$B_t = \max\{F_t, G\} \tag{5.1}$$

The quantity

$$K_t = B_t - F_t = \max\{G - F_t, 0\} \tag{5.2}$$

corresponds to the pay-off of a *put option*.

Alternative definitions of the guaranteed amount can be chosen so that the difference $B_t - F_t$ corresponds to the pay-off of a given financial option.

Remark When a financial guarantee is underwritten, a financial risk emerges for the insurer. Such a risk needs to be appropriately hedged, through a suitable asset management strategy. Hence, before underwriting a guarantee, the insurer must investigate whether it is possible to hedge it. Therefore, usually the insurer looks at hedging strategies available on the market, and then selects the guarantee offered to the policyholder.

For example, under the benefit

$$B_t = \max\{F_t, \max\{F_s\}_{s=0,1,\dots,t-1}\} \tag{5.3}$$

it is guaranteed that the minimum amount paid at time t is the highest value of the policy fund experienced at the previous policy anniversaries.

The guaranteed amount in this case is defined as follows:

$$G_t = \max\{F_s\}_{s=0,1,\dots,t-1} \tag{5.4}$$

Then:

$$K_t = B_t - F_t = \max\{\max\{F_s\}_{s=0,1,\dots,t-1} - F_t, 0\} \tag{5.5}$$

and this corresponds to the pay-off of a *ratchet option*.

With reference to the maturity benefit, the following guarantee, aimed at protecting the policyholder's investment, can be provided:

$$S_m = \max\{F_m, G_m\} \tag{5.6}$$

with

$$G_m = \sum_{t=0}^{m-1} P_t^{[S]} (1+i)^{m-t} \tag{5.7}$$

where $i \geq 0$ and $P_t^{[S]}$ denotes the part of premium invested in the unit-fund (the "savings premium" according to the traditional actuarial language).

It goes beyond the scope of this book to deal with issues related to asset management and asset valuation. Here we just provide a description of the structure of the financial options included in unit-linked policies.

Generalizing the guarantee expressed by Eq. (5.1), we assume that the benefit payable at time t (in case of death or, if $t = m$, in case of survival or death) is defined as follows:

$$B_t = \max\{F_t, G_t\} \tag{5.8}$$

where G_t is the guaranteed amount. Rearranging (5.8), the benefit can be expressed as follows:

$$B_t = F_t + \max\{G_t - F_t, 0\} \tag{5.9}$$

or as follows:

$$B_t = G_t + \max\{F_t - G_t, 0\} \tag{5.10}$$

According to (5.9), the benefit consists of the policy fund (whose value is unknown) and the pay-off of a *put option*, with strike G_t and underlying the reference fund. The maturity of the option is time t, the time of possible payment of the benefit. Conversely, according to (5.10) the benefit consists of a fixed amount G_t, like a traditional policy, to which the pay-off of a *call option* is added. The strike, the underlying and the maturity of the call option are clearly the same as those of the put option (given that we are describing the same benefit). For unit-linked policies, the description provided by (5.9) is more natural than (5.10), as the main feature of the arrangement is to realize an investment in the reference fund. However, when addressing the calculation of the cost of the guarantee, sometimes it is easier to assess the cost of a call option, and then reference would be made to (5.10).

According to the way the strike is defined, we can investigate further the structure of the option. If $G_t = G$, constant, the option is European-like, while if G_t depends on the past performances of the policy fund, such as in (5.4), the option is path-dependent. If G_t is a function of the premiums paid, the guarantee is endogenous. If guarantee (5.4) is chosen and a single premium was paid, the value of the option just depends on the investment performance (while, when it is endogenous, its value also depends on choices concerning the invested amount).

If a guarantee is underwritten in a unit-linked policy, a fee is required to the policyholder. In general, it is easier to assess the cost of a European-like option than that of a path dependent option, as well as it is easier to evaluate an option with exogenous guarantees than with endogenous guarantees. A market-consistent assessment is required, following the common practice for the pricing of financial derivatives. This requires a calibration to market data, even if the option is not directly traded on the market. Reference should be made to similar options. However, options traded in financial markets have many differences in respect of those included in life policies, such as the maturity (which is typically shorter for traded options), a different underlying, a different strike. Further, it should be noted that the exercise of options in (5.9) and (5.10) is subject not only to economic events (namely: it is convenient to exercise the call option if $F_t > G_t$, while it is convenient to exercise the put option if $F_t < G_t$), but also to the lifetime of the insured. Indeed, the benefit at time t is payable in case of death (or survival); this aspect adds complexity to the valuation of the insurer's liability.

As regards surrender guarantees, we note that they may be expressed similarly to (5.8); clearly, the benefit would be the surrender value R_t instead of B_t. The guaranteed amount G_t is usually defined so to provide a financial protection to the investment of the policyholder; therefore, the amount G_t typically implies a minimum (annual or average) return on the amount invested. The exercise of the surrender guarantee depends on economic events (the exercise is convenient if $G_t > F_t$), but also on preferences of the policyholder (whether to maintain or not the policy). This latter aspect is very hard to model; surrender guarantees represent important costs for insurers, but their assessment is still an open problem, due to the difficulty in representing individual preferences.

5.9 Variable Annuities

The expression *variable annuity* is used to refer to a wide range of life insurance products, whose benefits can be protected against financial and mortality/longevity risks by selecting one or more guarantees out of a broad set of possible arrangements.

Originally developed for providing a post-retirement income with some degree of flexibility, nowadays accumulation and death benefits constitute important components of the product design. Indeed, the variable annuity can be shaped so as to offer dynamic investment opportunities with some guarantees, benefit in the case of early death and a post-retirement income.

5.9.1 Main Features of the Products

We stress that no guarantee is implicitly embedded in a variable annuity product, whereas one or more guarantees can be chosen by the client and then added to the product. Hence, the guarantee structure is the result of the options exercised by the client (see Fig. 5.7). Guarantees are usually denoted by GMxB, that is Guaranteed Minimum Benefit of type x.

In what follows we assume, for simplicity, that the accumulation only relies on the payment of a single premium Π at the time the policy is written, i.e., time 0. The decumulation phase provides the insured with a post-retirement income, for example via a drawdown process. Conversely, we assume that no withdrawals occur prior to retirement time r. Let F_t denote the *policy account value* (or *policy fund value*) at time t, which depends on the evolution of the reference fund in which the single premium is invested.

Figures 5.8 and 5.9 show possible time profiles of the policy fund value in two different performance scenarios, that is, in the case of "good" performance (Scenario 1) and in the case of "bad" performance (Scenario 2), throughout both the accumulation and the decumulation phase. We assume absence of guarantees.

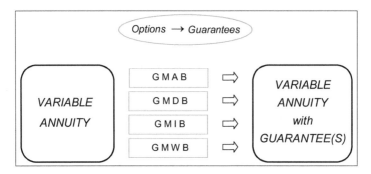

Fig. 5.7 Options and related guarantees in Variable Annuity products

Fig. 5.8 Time profile of the
policy account value
(Scenario1)

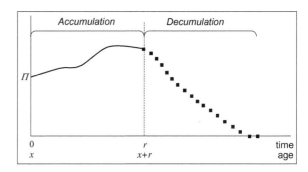

Fig. 5.9 Time profile of the
policy account value
(Scenario2)

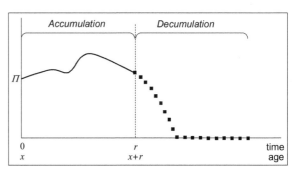

In the following sections we will see how the various GMxB can increase the fund available, in both the phases.

5.9.2 Guaranteed Minimum Accumulation Benefit (GMAB)

The GMAB is usually available prior to retirement. At some specified date, the insured (if alive) is credited the greater between the policy account value and a guaranteed amount. Assuming that the guarantee refers at retirement time r, the guaranteed amount, $G_r^{[A]}$, can be stated according to the following arrangements.

- *Return of premium*:

$$G_r^{[A]} = \Pi \tag{5.11}$$

- *Roll-up guarantee*:

$$G_r^{[A]} = \Pi\,(1+i)^r \tag{5.12}$$

  where i is the guaranteed interest rate.
- *Ratchet guarantee*:

$$G_r^{[A]} = \max_{t_h < r}\{F_{t_h}\} \tag{5.13}$$

where t_h, $h = 1, 2, \ldots$, are the stated ratchet times; hence, the profits of the reference fund are locked-in at the ratchet times, and then the guaranteed amount never decreases.

We note that, according to guarantees expressed by Eqs. (5.11) and (5.12), the guaranteed amount is fixed and hence known at the time the policy is written, whereas the guarantee expressed by Eq. (5.13) yields an amount which depends on the fund performance and is then unknown at time 0.

In principle, guarantees can be combined; for example:

- *Roll-up and Ratchet guarantee*:

$$G_r^{[A]} = \max\left\{\Pi\,(1+i)^r,\ \max_{t_h < r}\{F_{t_h}\}\right\} \tag{5.14}$$

As the result of any given guarantee mechanism, the amount $B_r^{[A]}$ acknowledged at time r is defined as follows:

$$B_r^{[A]} = \max\{F_r, G_r^{[A]}\} \tag{5.15}$$

Figures 5.10 and 5.11 illustrate the behavior of the policy account value (the solid line) in the two performance scenarios referred to in Figs. 5.8 and 5.9, as well as the time profile of the return of premium guarantee (see Eq. (5.11)) and the roll-up guarantee (Eq. (5.12)). In the case of good performance (Scenario 1), no guarantee is used, while in the case of bad performance (Scenario 2), if the roll-up guarantee has been chosen by the policyholder, then the consequent amount will be available at maturity.

Figures 5.12 and 5.13 show the effect (in Scenario 2) of ratchet guarantees (see Eq. (5.13)) with different time intervals.

Remark In the presence of a GMAB, a *Guaranteed Minimum Surrender Benefit* (*GMSB*) can be added. The amount of the guaranteed surrender benefit is usually determined consistently with the guaranteed accumulation benefit.

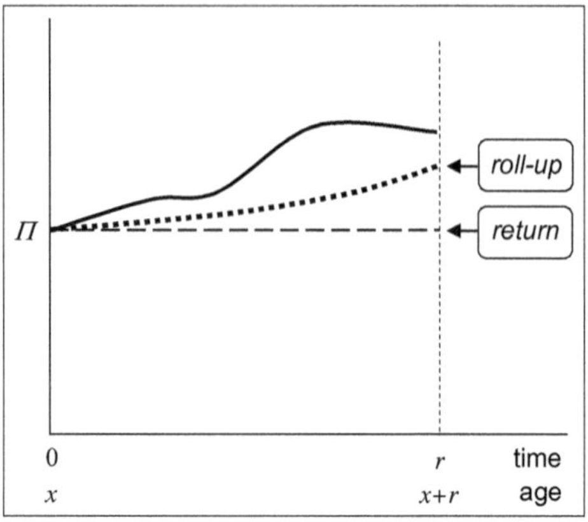

Fig. 5.10 GMAB—return
of premium and roll-up
guarantees (Scenario 1)

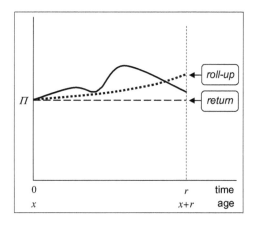

Fig. 5.11 GMAB—return
of premium and roll-up
guarantees (Scenario 2)

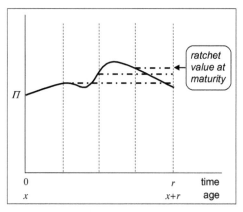

Fig. 5.12 GMAB—Ratchet
guarantee with "low"
frequency (Scenario 2)

Fig. 5.13 GMAB—Ratchet
guarantee with "high"
frequency (Scenario 2)

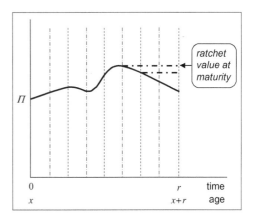

5.9.3 Guaranteed Minimum Death Benefit (GMDB)

Similarly to the GMAB, also the GMDB is available during the accumulation period; however, some insurers are willing to provide a GMDB also after retirement, up to some maximum age (say, 75 years). The structure of the guarantee is similar to the GMAB: in case of death at time t, prior to the stated term of the guarantee, the insurer will pay the greater between the policy account value and a stated amount $G_t^{[D]}$. Hence, the death benefit at time t is given by:

$$B_t^{[D]} = \max\{F_t, G_t^{[D]}\} \tag{5.16}$$

The guaranteed amount $G_t^{[D]}$ can be defined according to formulae similar to those adopted for the GMAB.

- *Return of premium*:

$$G_t^{[D]} = \Pi \tag{5.17}$$

- *Roll-up guarantee*:

$$G_t^{[D]} = \Pi (1+i)^t \tag{5.18}$$

- *Ratchet guarantee*:

$$G_t^{[D]} = \max_{t_h < t}\{F_{t_h}\} \tag{5.19}$$

where t_h, $h = 1, 2, \dots$, are the stated ratchet times.

Moreover, the following guarantee can be chosen:

- *Reset guarantee*:

$$G_t^{[D]} = F_{\max\{t_j : t_j < t\}} \tag{5.20}$$

where t_j, $j = 1, 2, \dots$, are the stated reset times.

Fig. 5.14 GMDB—return
of premium and roll-up
guarantees (Scenario 2)

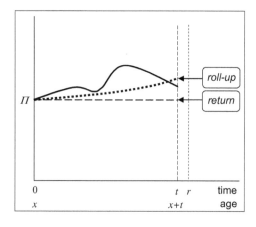

Fig. 5.15 GMDB—reset
guarantee (Scenario 2)

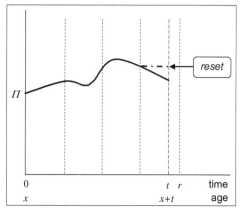

We note that the difference between the ratchet and the reset guarantee within the GMDB stands in the time profile of the guaranteed minimum amount: in the ratchet guarantee the minimum amount never decreases, while a reduction may occur in the reset guarantee, if the policy account value decreases between two reset times.

Also the GMDB can in principle be defined as a combination of guarantees, e.g., Roll-up and Ratchet (see Eq. (5.14), in which the maturity time r must be replaced by a generic time t).

Figures 5.14 and 5.15 show the effects (in Scenario 2) of three GMDB guarantees. In particular, Fig. 5.14 illustrates the return of premium and the roll-up guarantee; in the case of death at time t, if the roll-up guarantee has been chosen by the policyholder, then the amount $\Pi (1 + i)^t$ will be paid to the beneficiary. The effect of the reset guarantee is finally shown in Fig. 5.15.

5.9.4 Guaranteed Minimum Income Benefit (GMIB)

The GMIB provides the insured with a whole-life annuity from time r on. Let $b^{[I]}$ denote the guaranteed annual benefit. We focus on the following arrangements.

- *Guarantee on the amount to annuitize*:

$$b^{[I]} = \max\{F_r, G_r^{[I]}\} \frac{1}{\ddot{a}_{x+r}^{[\text{curr}]}} \tag{5.21}$$

where $x + r$ is the policyholder's age at time r, and $G_r^{[I]}$ can be defined as $G_r^{[A]}$ (see Eqs. (5.11)–(5.13)). Thus, the annuity rate is stated at time r, on the basis of current market conditions, and hence unknown up to that time.

- *Guarantee on the annuity rate* (stated before time r, in particular at the date the policy is issued):

$$b^{[I]} = F_r \max \left\{ \frac{1}{\ddot{a}_{x+r}^{[\text{curr}]}}, \frac{1}{\ddot{a}_{x+r}^{[\text{guar}]}} \right\} \tag{5.22}$$

This guarantee is also known as the *GAO* (*Guaranteed Annuity Option*).[2]

In principle, the two guarantees can be combined, with the following result:

- *Guarantee on the amount and the annuity rate*:

$$b^{[I]} = \max\{F_r, G_r^{[I]}\} \max \left\{ \frac{1}{\ddot{a}_{x+r}^{[\text{curr}]}}, \frac{1}{\ddot{a}_{x+r}^{[\text{guar}]}} \right\}$$

In practice, the resulting variable annuity product would be very expensive, because of the huge risk (which originates from both the fund performance and the longevity dynamics) taken by the insurer.

5.9.5 Guaranteed Minimum Withdrawal Benefit (GMWB)

The GMWB guarantees periodical withdrawals from the policy account, from time r on, even if the policy account value reduces to zero because of:

- poor investment performance;
- insured's long lifetime.

The guarantee affects both:

1. the withdrawal amount;

---

[2]As is well known, an accumulation-decumulation product with features quite similar to those expressed by the GAO caused the demise of the Equitable Life Assurance Society of London.

2. the withdrawal duration, which may be

   a. fixed, regardless of whether the retiree is alive or not;
   b. fixed, provided that the retiree is alive;
   c. lifelong.

In the case of guaranteed durations (a) and (b), if the policy account at the time of retiree's death is positive, then the amount is credited to the retiree's estate. In the case of guaranteed duration (c), the guarantee is also known as *Guaranteed Lifetime Withdrawal Benefit (GLWB)*.

The withdrawal amount, $b_t^{[W]}$, is stated as a given percentage, β_t (possibly constant), of a base amount W_t which is usually the policy account value at the date t^* the GMWB is selected. Hence:

$$b_t^{[W]} = \beta_t \, F_{t^*} \tag{5.23}$$

In Fig. 5.16 the time profile of the fund available under a GMWB of type (a) is displayed, versus the time profile of the policy account value. We assume $t^* = r$, and $\beta_t = 0.05$ for $t = r, r+1, \ldots$. It follows that the guaranteed annual withdrawal is 5% of the fund available at retirement, thus $b_t^{[W]} = 0.05 \, F_r$. Hence, withdrawals are guaranteed for 20 years, even in the case of fund depletion because of a poor performance, as shown in the figure.

In some arrangements, at specified dates (e.g., every policy anniversary) the base amount, W_t, may step up to the current value of the policy account, if this is higher. This is a ratchet guarantee, which may be lifetime or limited to some years (10 years, say). According to the ratchet mechanism, we have:

$$b_t^{[W]} = \beta_t \, W_t = \beta_t \, \max\{F_{t^*}, F_t\} \tag{5.24}$$

Fig. 5.16 GMWB—Withdrawals guaranteed for 20 years

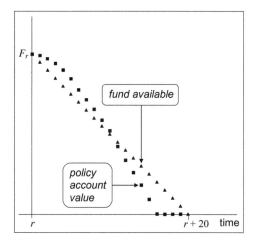

Remark The GMWB is the real novelty of variable annuities in respect of traditional life insurance and life annuity contracts. It provides a benefit which is similar to an income drawdown, but with guarantees. When comparing a GMIB to a GMWB, three major differences arise:

(i) the duration of the annuity (which is lifetime in the GMIB, while different conditions can be chosen in the GMWB as specified under point 2 above);
(ii) the accessibility to the account value (just for the GMWB);
(iii) the features of the reference fund and the consequent financial structure of the annuity, which usually is unit-linked in the GMWB, but typically participating in the GMIB.

It is also interesting to compare the GLWB, i.e., a GMWB with guaranteed withdrawal duration (c), with a standard life annuity. On the one hand, both the arrangements pay out lifelong benefits, but, on the other hand, the higher flexibility of the GMWB (consider, in particular, point (ii) above, but also the unit-linked financial structure) implies, under the same market conditions, a higher cost to the insurer and hence, for a given single premium, an (initial) benefit in the GMWB arrangement smaller than that provided by the standard life annuity.

5.9.6 Risks in Variable Annuities

Risks borne by an insurance company selling variable annuities of course depend on the options exercised by the policyholders, and hence on the guarantees provided by the contracts.

Basically, we can single-out the following classes of risks causes:

1. financial risks, originated by minimum guarantees on the performance of investments in units/equities;
2. biometric risks, and in particular:

   a. mortality risk, depending on minimum guarantees on death benefits;
   b. longevity risk, depending on either minimum guarantees on withdrawals or by features of life annuities.

We note that, according to the structure of the guarantee, more than one risk cause can impact on the risk profile of the product and then of the portfolio. Consider the following examples.

- According to the GMDB, a death benefit will be paid to the beneficiary; a mortality risk then follows. The amount of the benefit has $G_t^{[D]}$ as the floor (see Eq. (5.16)). We note that, unlike the "ordinary" term insurance (see Sect. 5.3), a significant financial risk is also borne by the insurer because of the randomness inherent in the fund value.
- The GMIB defined by (5.21) implies during the accumulation period a financial risk quite similar to that inherent in the GMAB. If the annuitization option is exercised,

the GMIB generates a life annuity. Then, the same risks which feature an immediate life annuity (see Sect. 5.5) are borne by the insurer during the decumulation period, provided that the reserve is invested in bonds rather than in units.

- Conversely, the GMIB defined by (5.22) does not imply a financial risk during the accumulation period, but, with respect to the GMIB defined by (5.21), implies a much higher amount of longevity risk, depending on the time when the annuity rate is guaranteed.
- Risks implied by the GMWB depend on the specific type of guarantee on the withdrawal duration chosen by the client (see point 2 in Sect. 5.9.5). Focusing on the decumulation period, we note that, whatever specific guarantee has been chosen, a financial risk is implied. Longevity risk is, in particular, the consequence of the guaranteed lifelong withdrawal duration, that is the GLWB (see also Remark in Sect. 5.9.5).
- Possible withdrawals during the accumulation period (not considered in the previous sections) as well as surrender might cause liquidity risk. Further, the presence of a GMSB (see Remark in Sect. 5.9.2) also implies financial risk.

A variable annuity can of course include more than one GMxB. For example, the client can choose the GMDB which works over the whole accumulation period (with possible extension to the initial years of the decumulation period), and the GMIB defined by (5.21), which will provide a life annuity as post-retirement income. It follows that financial risk, mortality risk and longevity risk will be borne by the insurance company.

5.10   References and Suggestions for Further Reading

A wide range of life insurance and life annuity products and related guarantees and options are described by Black and Skipper (2000), where also health insurance products are addressed.

The actuarial structure of "basic products" (e.g., term insurance, endowment insurance, life annuities) is described in all the textbooks dealing with life insurance mathematics and technique; see, for example Olivieri and Pitacco (2015) and references therein.

Risk classification in life insurance and life annuities is addressed in many books and papers; a compact review, together with an extensive reference list, is provided by Haberman and Olivieri (2014).

A number of recent papers focus on participation mechanisms in life insurance. See, for example, Reuss et al. (2015, 2016) (and the extended lists of references therein), Gatzert et al. (2012a), Heling and Holder (2013), and Maurer et al. (2013).

A world-wide perspective of underwriting procedures is provided by Klein (2018). Preferred risks in life insurance are dealt with by Hughes (2012).

Life annuities and related guarantees and options are addressed, for example, by Pitacco (2016a, 2017). Life annuities in the framework of retirement income strategies are described by Milevsky (2013). Mortality trends and the relevant impact on pension funds and life annuity portfolios are dealt with in Pitacco et al. (2009), where RM issues are also presented. Guaranteed annuity options (GAO) are analyzed by Boyle and Hardy (2003).

Underwritten life annuities (or special-rate life annuities) are described in various papers and technical reports: see, in particular, Ainslie (2000), Drinkwater et al. (2006), O'Brien (2013), Ridsdale (2012), and Rinke (2002). The article by Edwards (2008) is specifically devoted to life annuity rating based on postcodes. The use of postcodes allows to express the age-pattern of mortality as a function of the annuitant's social class and geographic location of the housing.

The impact of risk classification on the structure of life annuity portfolios is dealt with by Gatzert et al. (2012b), Hoermann and Russ (2008) and Olivieri and Pitacco (2016). An extensive literature focuses on the impact of heterogeneity due to unobservable risk factors, usually summarized by the individual "frailty", on the results of a life annuity portfolio. For a detailed bibliography, the reader can refer to Pitacco (2019), where relations between mortality at high ages and frailty are also addressed.

The unisex rating principle in insurance products, life annuities and pensions is addressed by recent papers. See, for example: Chan (2014), Curry and O'Connell (2004), and Oxera Consulting (2010).

Hedging longevity risk in life annuity business via appropriate product design is the topic of Olivieri and Pitacco (2020), where also an extensive list of references can be found.

Long-term care insurance products are described, for example, in Pitacco (2014), where an extensive list of references is also provided. An updated perspective on various aspects of long-term care and long-term care insurance is given by IAA (2017). Disability severity measures and the OPCS scale in particular are addressed by Martin and Elliot (1992).

The book by Hardy (2003) focuses on investment risks and relevant guarantees in equity-linked life insurance products. Kalberer and Ravindran (2009) and Ledlie et al. (2008) specifically deal with variable annuities and related guarantees and options.

Actuarial involvement in all the phases of the RM process calls for a sound awareness of insurance management problems, even beyond the traditional technical aspects (e.g., pricing, reserving, reinsurance, etc.). Marketing approach to the product design, choice of distribution channels and analysis of policyholders' behaviour are significant examples. A comprehensive view on the management of life annuity business, with specific reference to the UK market is provided by Telford et al. (2011). Two papers in the IAA Risk Book specifically deal with two of the above problems, referring to both life and non-life lines of business: Gutterman (2016) focuses on risks inherent in the distribution of insurance products, while Niittuinperä (2020) addresses various aspects of the policyholder's behaviour.

Papers and reports addressing product development, product design and related technical issues will be cited in Sect. 9.4.

Chapter 6
Risk Assessment and Impact Assessment in Life Insurance Business

6.1 A Stochastic Model for a Life Insurance Portfolio

In this section, we define all the quantities needed to construct a specific model, in the framework presented in Sect. 4.4.3. We then illustrate various implementations of the model, aiming at impact assessment.

6.1.1 The Portfolio

We refer to a portfolio of participating endowment insurance policies , consisting of one generation; all policies are issued at time $t = 0$ with term m years; x_0 denotes the age at policy issue.

According to policy conditions initially stated (that is, at the policy issue), the following quantities are defined:

$C =$ lump sum benefit in the case of death in year t, i.e. between $t - 1$ and t ($t = 1, 2, \ldots, m$), for simplicity assumed payable at time t, or in the case of survival at maturity (time m);
$P =$ annual net premium;
$P^{[T]} =$ annual gross premium;
$V_t =$ policy reserve at time t ($t = 0, 1, \ldots, m$);
$R_t =$ surrender value at time t ($t = 1, 2, \ldots, m - 1$), as a function of the policy reserve; usually $R_1 = R_2 = 0$.

We assume that premiums are payable for the whole policy duration, that is, at times $t = 0, 1, \ldots, m - 1$. Premiums and (net) policy reserves are calculated, according to the equivalence principle, as follows (the traditional actuarial notation is adopted).

All the examples in Sect. 6.1 have been planned and elaborated by Daniela Y. Tabakova.

© Springer Nature Switzerland AG 2020
E. Pitacco, *ERM and QRM in Life Insurance*, Springer Actuarial,
https://doi.org/10.1007/978-3-030-49852-8_6

- Net premium:

$$P = \frac{C \, A_{x_0,\overline{m}|}}{\ddot{a}_{x_0:\overline{m}|}} \tag{6.1}$$

- Gross premium:

$$P^{[T]} = \frac{C \left(\frac{A_{x_0,\overline{m}|}}{\ddot{a}_{x_0:\overline{m}|}} + \gamma \right)}{1 - \beta - \frac{\delta(m)}{\ddot{a}_{x_0:\overline{m}|}}} \tag{6.2}$$

where the expense loading parameters are defined as follows:

β = loading for premium collection expenses as a percentage of the gross premium;
γ = loading for annual general expenses, as a percentage of the benefit;
$\delta(m)$ = loading for acquisition expenses, as a percentage of the gross premium.

- Policy reserve at time t:

$$V_t = C \, A_{x_0+t,\overline{m-t}|} - P \, \ddot{a}_{x_0+t:\overline{m-t}|} \tag{6.3}$$

- Zillmer policy reserve at time t:

$$V_t^{[\text{Zill}]} = V_t - \frac{\delta(m) \, P^{[T]}}{\ddot{a}_{x_0:\overline{m}|}} \ddot{a}_{x_0+t:\overline{m-t}|} \tag{6.4}$$

- Surrender value at time t:

$$R_t = \begin{cases} 0 & \text{for } t = 1, 2 \\ \rho \, V_t^{[\text{Zill}]} & \text{for } t = 3, \ldots, m - 1 \end{cases} \tag{6.5}$$

with $0 < \rho \leq 1$.

We assume that a participation mechanism works, which is based on the investment extra-return with respect to the "technical" rate of interest, i, adopted in premium and reserve calculations. The participation mechanism works as follows.

- $j_t^{[\Pi]}$, $j_t^{[V]}$, $j_t^{[B]}$ denote the rates of adjustment at time t of the premium, the reserve and the benefit, respectively;
- The annual rate of increase of the reserve is defined, according to the traditional participation mechanism as follows:

$$\max\{i, \eta_t \, g_t\}; \quad t = 1, 2, \ldots \tag{6.6}$$

It can be proved that the rate of adjustment of the reserve, $j_t^{[V]}$, also named "revaluation" rate and shortly denoted by r_t, is then given by:

$$r_t = j_t^{[V]} = \max\left\{\frac{\eta_t\, g_t - i}{1 + i}, 0\right\} \tag{6.7}$$

where:

i = technical rate of interest;
η_t = participation share in year t;
g_t = investment yield in year t.

- The rate of adjustment of the premium, $j_t^{[\Pi]}$, is usually linked to the reserve revaluation rate, or set to 0.
- It can be proved that the following relation holds:

$$j_t^{[B]} = \frac{j_t^{[V]}\, V_{t^-} + j_t^{[\Pi]}\, P_{t-1}\, \ddot{a}_{x_0+t:\overline{m-t}\rceil}}{V_{t^-} + P_{t-1}\, \ddot{a}_{x_0+t:\overline{m-t}\rceil}} \tag{6.8}$$

where V_{t^-} denotes the policy reserve at time t before the adjustment at that time:

$$V_{t^-} = C_t\, A_{x_0+t,\overline{m-t}\rceil} - P_{t-1}\, \ddot{a}_{x_0+t:\overline{m-t}\rceil} \tag{6.9}$$

- Policy reserves, premiums and benefits evolve as follows:

$$V_t = V_{t^-}\,(1 + j_t^{[V]}); \quad t = 1, 2, \ldots, m \tag{6.10}$$
$$P_t = P_{t-1}\,(1 + j_t^{[\Pi]}); \quad t = 1, 2, \ldots, m - 1, \text{ with } P_0 = P \tag{6.11}$$
$$P_t^{[T]} = P_{t-1}^{[T]}\,(1 + j_t^{[\Pi]}); \quad t = 1, 2, \ldots, m - 1, \text{ with } P_0^{[T]} = P^{[T]} \tag{6.12}$$
$$C_t = C_{t-1}\,(1 + j_{t-1}^{[B]}); \quad t = 2, 3, \ldots, m, \text{ with } C_1 = C \tag{6.13}$$

Finally, the benefit at maturity is given by:

$$C_m^* = C_m\,(1 + j_m^{[B]}) = C_m\,(1 + j_m^{[V]}) \tag{6.14}$$

At time 0, the following amounts are random because of future profit participation over the policy duration:

C_t = benefit in the case of death in year t, payable at time t ($t = 2, \ldots, m$);
C_m^* = benefit in the case of survival at maturity;
P_t = net annual premium at time t ($t = 1, 2, \ldots, m - 1$);
$P_t^{[T]}$ = gross annual premium at time t ($t = 1, 2, \ldots, m - 1$);
V_t = policy reserve at time t ($t = 1, 2, \ldots, m$);
EX_t = expenses at time t ($t = 1, 2, \ldots, m - 1$);
R_t = surrender value at time t ($t = 1, 2, \ldots, m - 1$).

The following random quantities are defined at portfolio level:

N_t = number of policies in-force at time t, with initial number N_0 given;
D_t = number of deaths in year t, i.e. between time $t - 1$ and t;

A_t = number of lapses/surrenders in year t, assumed to occur at time t ($t =$ $1, 2, \ldots, m - 1$).

Then:

$$N_t = N_{t-1} - D_t - A_t \tag{6.15}$$

Hence, at portfolio level we have:

$C_t^{[P]} = C_t\, D_t$ = total amount of death benefits paid at time t, i.e. because of deaths between $t - 1$ and t;
$P_t^{[T][P]} = P_t^{[T]}\, N_t$ = total amount of gross premiums cashed at time t;
$EX_t^{[P]} = EX_t\, N_t$ = total amount of expenses paid at time t;
$C_m^{*\,[P]} = C_m^{*}\, N_m$ = total amount of benefits paid at maturity;
$R_t^{[P]} = R_t\, A_t$ = total amount of surrender benefits paid at time t:
$V_t^{[P]} = V_t\, N_t$ = portfolio reserve at time t;
$V_t^{[Zill][P]} = V_t^{[Zill]}\, N_t$ = Zillmer portfolio reserve at time t.

Moreover, we define the following random quantities:

$F_t^{[P]}$ = portfolio fund (also named the "life fund") at time t; see Eqs. (6.18);
I_t = investment rate of return in year t;
M_t = surplus, that is shareholders' capital (assigned to the portfolio) at time t:

$$M_t = F_t^{[P]} - V_t^{[P]} \tag{6.16}$$

We assume absence of shareholders' capital flows (allocations or releases), besides (possible) initial allocation which leads to:

$$F_0^{[P]} = M_0 \tag{6.17}$$

Remark 6.1 We note that the portfolio reserve has simply been defined as $V_t^{[P]} = V_t\, N_t$, where V_t denotes the individual policy reserve calculated by adopting a prudential technical basis. Thus, for simplicity, we have disregarded the calculation of the portfolio reserve as the best-estimate reserve plus the risk margin. Although the portfolio reserve defined by the product of the policy reserve times the number of policies is consistent with various local GAAPs, it should be stressed that it does not take into account diversification effects.

The portfolio fund dynamics is defined as follows:

$$F_t^{[P]} = (F_{t-1}^{[P]} + P_{t-1}^{[T][P]} - EX_{t-1}^{[P]})\,(1 + I_t) - C_t^{[P]} - R_t^{[P]}; \quad t = 1, 2, \ldots, m - 1 \tag{6.18a}$$

$$F_m^{[P]} = (F_{m-1}^{[P]} + P_{m-1}^{[T][P]} - EX_{m-1}^{[P]})\,(1 + I_m) - C_m^{[P]} - C_m^{*\,[P]} \tag{6.18b}$$

From the first recursion we obtain:

$$F_t^{[P]} = M_0 \prod_{h=1}^{t}(1 + I_h) + \sum_{h=1}^{t}\left[(P_{h-1}^{[T][P]} - EX_{h-1}^{[P]})(1 + I_h) - (C_h^{[P]} + R_h^{[P]})\right] \prod_{j=h+1}^{t}(1 + I_j)$$

$$(6.19)$$

A similar result can be derived from the second recursion.

The fund dynamics described by Eqs. (6.18) and (6.19) disregards, as already noted, shareholders' capital flows, that is, capital allocation policies (besides the initial allocation M_0). Of course, a more realistic modeling should also allow for both capital releases (when $F_t^{[P]}$ exceeds a stated upper barrier) and further capital injections (if $F_t^{[P]}$ drops below a stated lower barrier). This simplification should be taken into account while interpreting the numerical examples in the following sections, which only provide a first insight into the analysis of the portfolio fund behavior.

Remark 6.2 Equation (6.19) provides an example of *aggregation function* (see Sect. 4.4.3); in particular:

- $F_t^{[P]}$ is the (annual) result of interest;
- capital allocation M_0 is a decision variable;
- random yields I_h are scenario variables;
- the premiums $P_{h-1}^{[T][P]}$ depend on

  - scenario variables, e.g., previous yields I_j, number of deaths D_j, etc.;
  - decision variables which constitute the first-order basis to calculate $P_0^{[T]}$;

- ......

6.1.2 From Theory to Practice

The examples in the following sections show basic applications of QRM to a life insurance portfolio, in line with the approaches described in Sect. 4.4. In particular:

- The structure of a specific endowment insurance portfolio is defined in Example 6.1; technical bases, premiums and reserves are provided.
- Various scenarios are defined in Example 6.2; diverse features allow us to perform several assessments and to compare the relevant results.
- Following the approaches described in Sect. 4.4.4, we present:

  - results of a deterministic analysis in Examples 6.3, 6.4, 6.5 and 6.6;
  - results of a stochastic analysis, only allowing for random fluctuations, in Examples 6.7 and 6.8;
  - results obtained by allowing for uncertainty, and hence possible systematic deviations, in Example 6.9;
  - results achieved by implementing a double-stochastic approach in Example 6.10.

- Finally, a simple analysis of the impact of a decision variable is presented in Example 6.11.

6.1.3 Defining an Endowment Portfolio

In the following example, the endowment portfolio is defined, which will be used in the risk assessment and impact assessment examples.

Example 6.1 We refer to the portfolio described in Sect. 6.1.1, and assume the following data.

- Initial portfolio size: $N_0 = 10\,000$;
- Benefit amount, i.e. sum assured, stated at policy issue: $C = 1\,000$;
- Initial age: $x_0 = 50$;
- Policy term: $m = 15$;
- Technical basis adopted in premium and reserve calculations, that is, the first-order technical basis:

  – interest rate: $i = 0.02$;
  – mortality assumption: first Heligman–Pollard law with the parameter values and markers specified in Example 4.1;
  – expenses are attributed to each policy according to the following loading parameters:

    · acquisition expenses: $\delta(15) = 50\%$ of the gross premium;
    · collection expenses: $\beta = 5\%$ of the gross premium per year;
    · general expenses: $\gamma = 1\%_o$ of the sum insured per year.

  Expenses attributed to each policy are then initially determined, for the first year and the following years respectively, as follows:

$$EX_0 = \left(\delta(m) + \beta\right) P^{[T]} + \gamma\, C \tag{6.20}$$

$$EX_t = \beta\, P^{[T]} + \gamma\, C; \quad t = 1, 2, \ldots, m - 1 \tag{6.21}$$

We then find the following initial annual premiums:

- net premium: $P = 59.54$;
- gross premium: $P^{[T]} = 66.50$.

Surrender values are defined by Eq. (6.5), with $\rho = 0.90$. As regards the participation mechanism, we assume:

- profit participation share $\eta = 95\%$ ($=$ const.);
- $j_t^{[\Pi]} = 0$ for all t, thus constant annual premiums.

∎

6.1.4 Scenarios

Several hypotheses are assumed in order to define possible scenarios. The scenario features (i.e. the risk causes) specified in Example 6.2 must in particular be considered, as we will see in the following examples in which hypotheses are implemented consistently with the approach (either deterministic or stochastic) adopted in the risk assessment and impact assessment phases.

Example 6.2 We consider the following features.

- *Investment yield.* Table 6.1 shows three hypotheses about the investment yield, assumed deterministic and constant over the whole policy duration. The scenarios are denoted by I_1, I_2, I_3, respectively. Alternatively, we can assume random investment yields, with an appropriate probabilistic structure. In particular, we assume that the value, $B(t)$, of the portfolio assets follows a geometric Brownian motion, that is, a stochastic process that satisfies the stochastic differential equation:

$$dB(t) = \mu B(t) \, dt + \sigma B(t) \, dW(t) \qquad (6.22)$$

where μ is the drift parameter, σ the volatility parameter, and $W(t)$ is a Wiener process. The solution of Eq. (6.22) is:

$$B(t) = B(0) \, e^{\left(\mu - \frac{\sigma^2}{2}\right)t + \sigma W(t)} \qquad (6.23)$$

from which, via simulation of the process $W(t)$, the outcomes of the annual yield can be derived. The following parameters are adopted in numerical examples:

$$\mu = \ln(1.04); \quad \sigma = 0.005 \qquad (6.24)$$

Moreover, three different investment scenarios with random yields can be defined by assuming:
$$\mu = \ln(1 + g^{[I_j]}); \quad j = 1, 2, 3 \qquad (6.25)$$

(see Table 6.1).
- *Mortality.* We describe mortality scenarios by defining the age-pattern of mortality in each scenario in terms of the first-order mortality, q_x, i.e. the mortality assumption adopted in premium and reserve calculations:

Table 6.1 Investment yield

I_j	$g^{[I_j]}$
I_1	3.0%
I_2	4.0%
I_3	5.0%

$$q_x^{[\text{scen}]} = \alpha\, q_x \quad \text{for all } x \tag{6.26}$$

As the first-order mortality, q_x, constitutes a safe-side assessment of insureds' mortality, Eq. (6.26) with $\alpha < 1$ can represent realistic assessments. Table 6.2 shows three mortality scenarios, denoted by D_1, D_2, D_3, respectively. Conversely, $\alpha > 1$ can be used to represent stress scenarios.

- *Expenses.* We adopt just one expense scenario, denoted by E. The expenses per policy are given for the first year by:

$$EX_0^{[E]} = \left(\delta^{[E]}(15) + \beta^{[E]}\right) P_0^{[T]} + \gamma^{[E]} C_1 \tag{6.27}$$

and for the following years by:

$$EX_t^{[E]} = \beta^{[E]} P_t^{[T]} + \gamma_t^{[E]} C_{t+1} \tag{6.28}$$

The portfolio expenses are then given for the first year by:

$$EX_0^{[P]} = EX_0^{[E]} N_0 \tag{6.29}$$

and for the following years by:

$$EX_t^{[P]} = EX_t^{[E]} N_t \tag{6.30}$$

We note that, while $EX_0^{[P]}$ is a deterministic amount, all the $EX_t^{[P]}$, for $t = 1, 2, \ldots, m - 1$, are random amounts because of the random numbers of in-force policies, N_t, and the participation mechanism which affects the sum insured. The following parameter values have been assumed:

$$\delta^{[E]}(15) = \delta(15) = 50\% \tag{6.31a}$$
$$\beta^{[E]} = 4.5\% \tag{6.31b}$$
$$\gamma_t^{[E]} = (0.8 + 0.02\,t)\,\%_o; \quad t = 0, 1, \ldots, m - 1 \tag{6.31c}$$

- *Lapses/Surrenders.* Three hypotheses, denoted by A_1, A_2, A_3, for the annual probabilities of abandoning the policy, $r_t^{[A_k]}$, are shown in Table 6.3. As we assume a zero surrender value for $t = 1, 2$, in each scenario $r_1^{[A_k]}$ and $r_2^{[A_k]}$ represent lapse probabilities, whereas the following $r_t^{[A_k]}$ represent surrender probabilities, $k = 1, 2, 3$.
- *Combining hypotheses.* Let S_{jhk} denote the generic scenario obtained by combining three hypotheses, each one concerning a risk cause:

$$S_{jhk} = I_j \wedge D_h \wedge A_k; \quad j, h, k = 1, 2, 3 \tag{6.32}$$

Table 6.2 Mortality

D_h	$q^{[D_h]}$
D_1	$0.70\,q$
D_2	$0.80\,q$
D_3	$0.90\,q$

In particular, we assume that $S_{222} = I_2 \wedge D_2 \wedge A_2$ describes the *best-estimate sce-nario*, that is, the most likely scenario. ■

6.1.5 Deterministic Approach

The simplest implementation of the deterministic approach consists in a single set of assumptions, which determine the output results of interest. Values must then be chosen for all the input variables (investment yield, mortality, surrenders). Possible choices are, for example, as follows:

- best-estimate values, in particular expected values based on the best-estimate prob-abilities, e.g., the best-estimate mortality assumption; see Example 6.3;
- best-case (i.e. optimistic) values, e.g., a very low mortality;
- worst-case (i.e. pessimistic values), e.g., a very high mortality.

The deterministic approach does not allow us to explore some important aspects. In particular:

1. what might be the possible impact of random fluctuations?
2. do the best-estimate values provide an appropriate representation of the reality, or should we account for possible systematic deviations?

Question 1 can only find answers in the framework of stochastic approaches. As regards question 2, we note that implementations of the deterministic approach, based on best-case and worst-case scenarios respectively, allow us to have a rough assessment of the impact of possible systematic deviations from best-estimate input values. In particular, *stress tests* can be performed by assigning worst-case values to one or more input variables; see Example 6.4.

The impact of different scenarios can be tested via iterative implementation of the deterministic approach. See Example 6.5.

Example 6.3 The path of the surplus, M_t, defined by Eq. (6.16), is shown in Figs. 6.1 and 6.2. In both the cases, the best-estimate scenario S_{222} has been assumed. Cal-culations have been performed by using Eq. (6.18) and replacing random quantities with expected values or estimates. In particular:

Table 6.3 Lapses/surrenders

A_k	$r_1^{[A_k]}$	$r_2^{[A_k]}$	$r_3^{[A_k]}$	$r_4^{[A_k]}$	$r_5^{[A_k]}$	$r_6^{[A_k]}$	$r_7^{[A_k]}$	$r_8^{[A_k]}$	$r_9^{[A_k]}$	$r_{10}^{[A_k]}$	$r_{11}^{[A_k]}$	$r_{12}^{[A_k]}$	$r_{13}^{[A_k]}$	$r_{14}^{[A_k]}$	$r_{15}^{[A_k]}$
A_1	3.0%	1.5%	3.0%	2.5%	2.0%	1.5%	1.0%	0.5%	0%	0%	0%	0%	0%	0%	0%
A_2	4.0%	2.5%	4.0%	3.5%	3.0%	2.5%	2.0%	1.5%	1.0%	0.5%	0.5%	0.5%	0.5%	0%	0%
A_3	5.0%	3.5%	5.0%	4.5%	4.0%	3.5%	3.0%	2.5%	2.0%	1.5%	1.5%	1.5%	1.5%	0%	0%

Fig. 6.1 Surplus path in the best-estimate scenario; $F_0^{[P]} = 0$

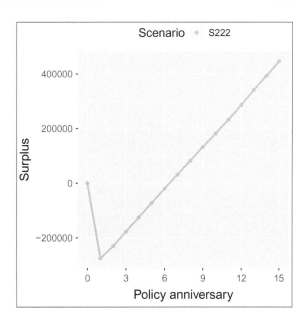

- random investment yield has been replaced by the best-estimate scenario I_2, that is, 4% (see Table 6.1);
- expenses have been estimated as specified by Eqs. (6.31), (6.29) and (6.30);
- expected numbers of deaths and expected numbers of lapses/surrenders have been determined according to the best-estimate probabilities D_2 and A_2 respectively (see Tables 6.2 and 6.3); the following relations hold among expected values, for $t = 1, 2, \ldots, m$:

$$\mathbb{E}[N_t] = \mathbb{E}[N_{t-1}] - \mathbb{E}[D_t] - \mathbb{E}[A_t] \quad (\text{with } \mathbb{E}[N_0] = N_0) \tag{6.33}$$

where:

$$\mathbb{E}[D_t] = \mathbb{E}[N_{t-1}] \, q_{x_0+t-1}^{[D_2]} \tag{6.34}$$

$$\mathbb{E}[A_t] = \mathbb{E}[N_{t-1}] \, (1 - q_{x_0+t-1}^{[D_2]}) \, r_t^{[A_2]} \tag{6.35}$$

The negative value of M_1 in Fig. 6.1 is caused by the new business strain, that is by the amount $E X_0^{[P]}$ which includes the initial expenses. A shareholders' capital allocation is then needed, to obtain $M_1 \geq 0$. In all the following examples, we assume the initial capital allocation which leads to $M_1 = 0$; see Fig. 6.2. ∎

Example 6.4 The impact of extreme mortality on the surplus path, via stress testing, is shown in Figs. 6.3 and 6.4. We assume, in terms of Eq. (6.26), $\alpha = 4$. In Fig. 6.3 the extreme mortality is assumed to occur from $t = 2$ onwards, and from $t = 12$ onwards in Fig. 6.4. We note that, even in the first case, where the impact is of course

Fig. 6.2 Surplus path in the
best-estimate scenario;
$F_0^{[P]} = 264\,406$

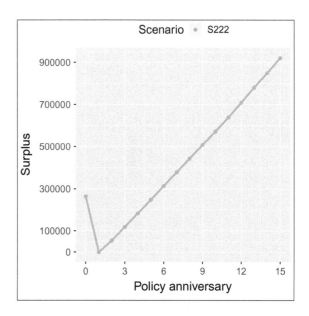

Fig. 6.3 Surplus path in the
best-estimate scenario and in
extreme mortality scenario
from $t = 2$

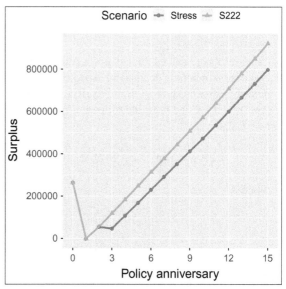

Fig. 6.4 Surplus path in the best-estimate scenario and in extreme mortality scenario from $t = 12$

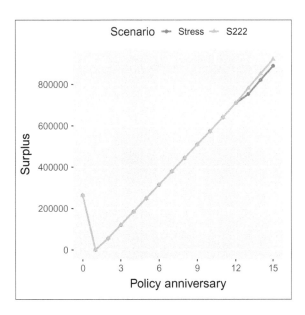

Fig. 6.5 Surplus path in three different scenarios (1)

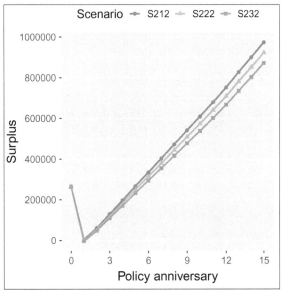

much more significant (and to some extent unrealistic), the reduction of the surplus is not dramatic: this is due to the type of insurance product, the endowment insurance, which is not heavily exposed to the mortality risk. ∎

Example 6.5 The impacts of three different scenarios are sketched in Figs. 6.5, 6.6, 6.7, 6.8 and 6.9. In all the cases, the impact of the best-estimate scenario S_{222} is

Fig. 6.6 Surplus path in
three different scenarios (2)

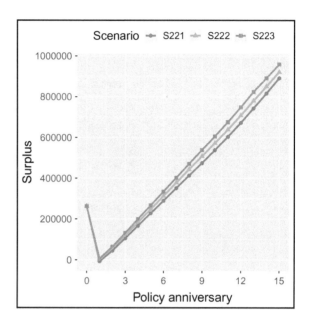

taken as a "benchmark". Alternative scenarios are denoted according to (6.32). We
note that the most significant diversity occurs when assumptions about investment
yield are involved, although the diversity in assumed yields is not dramatic (see
Table 6.1), as we can see from Figs. 6.7, 6.8 and 6.9. Again, this is due to the type
of insurance product, the endowment insurance which, because of the significant
financial component, is heavily exposed to the investment risk. ∎

Example 6.6 In the framework of sensitivity testing, *tornado plots* (or *tornado
charts*, or *tornado diagrams*) can help in summarizing the degree of impact each
input variable has on results of interest. The (negative or positive) changes, with
respect to the best-estimate scenario S_{222}, in surplus at time 15 as a consequence of
diverse investment (I), mortality (D) and lapses/surrenders (A) scenarios are shown
in Fig. 6.10. In particular:

- the top bar (I) refers to scenarios S_{122} and S_{322} (see Fig. 6.7);
- the middle bar (D) refers to scenarios S_{232} and S_{212} (see Fig. 6.5);
- the bottom bar (A) refers to scenarios S_{221} and S_{223} (see Fig. 6.6).

∎

Remark Examples 6.3 and 6.4 show the implementation of the deterministic
approach 1a (see Sect. 4.4.4): given the values of the input variables in each year
of the policy duration, the value of the output variable is calculated. Conversely,
Examples 6.5 and 6.6 illustrate the approach 1b: the impact of alternative values of
one of the input variable is tested in each implementation.

Fig. 6.7 Surplus path in
three different scenarios (3)

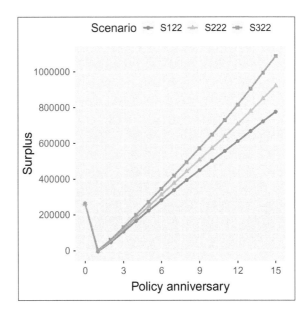

Fig. 6.8 Surplus path in
three different scenarios (4)

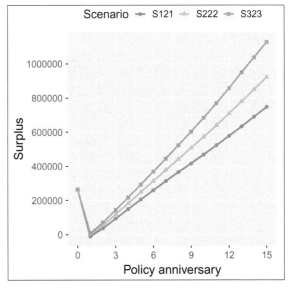

Fig. 6.9 Surplus path in
three different scenarios (5)

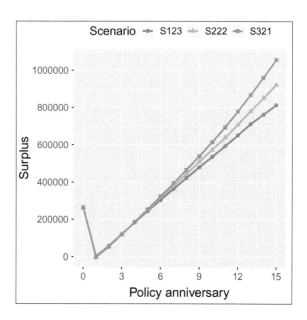

Fig. 6.10 Change in surplus
at time 15 as a consequence
of diverse scenarios

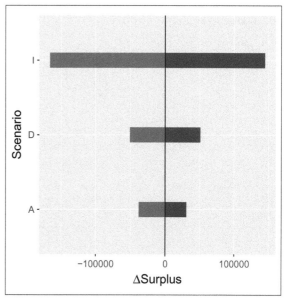

6.1.6   Stochastic Approach

The stochastic approach aims at the construction of probability distributions of results of interest, given the probability distributions of the input variables, and possible correlation assumptions. We note that, in the following examples, we assume stochastic independence among the input random variables (investment yield, mortality, lapses/surrenders). As calculations are performed via stochastic simulation, the distributions obtained for the results are empirical distributions (rather than probability distributions).

Stochastic simulation techniques adopted in the examples are as follows.

1. Random investment yields are simulated assuming the geometric Brownian motion model (see Eqs. (6.22) and (6.23)), with the chosen parameters (see Eqs. (6.24) and (6.25)).
2. Assuming stochastic independence among the individual random lifetimes, annual number of deaths are binomially distributed, with parameters given by the number of in-force policies at the beginning of the relevant year and the annual probability of death $q_x^{[\text{scen}]}$ corresponding to the relevant age x and the chosen scenario, e.g., the best-estimate scenario. An alternative to the simulation of the binomial is given by the simulation of the Poisson distribution with appropriate parameters.
3. Assuming stochastic independence among the individual lapse/surrender decisions, annual number of lapses/surrenders are binomially distributed, with parameters given by the number of in-force policies at the end of the relevant year and the annual probability of lapse/surrender corresponding to the chosen scenario, e.g., the best-estimate scenario. Again, an alternative to the simulation of the binomial is given by the simulation of the Poisson distribution with appropriate parameters.

Simulation procedures can be implemented in two distinct ways:

- *joint simulation* consists in the simulation of all the input variables, to obtain the corresponding distributions of the output variables;
- *marginal simulations* consist in separately simulating each input variable, so that the relevant impact on the randomness of the output variables can be assessed.

It is worth noting that, from a RM perspective, the joint simulation is of course needed to obtain, for example, the distribution of the surplus and then to determine capital requirements via VaR or TailVar. Conversely, marginal simulations allow us to single-out the impact of each risk cause, and hence, by comparing the different impacts, to take appropriate RM actions. For example, if investment risk appears the most significant, a new product design implying weaker financial guarantees can constitute an appropriate action. This particular problem will be dealt with in Sect. 6.2.

Example 6.7 Some results of joint simulations are sketched in Figs. 6.11, 6.12, 6.13, 6.14, and 6.15. In each figure, n_{sim} denotes the number of simulations. In Fig. 6.11, simulated surplus paths are sketched. Simulations have been performed according

Fig. 6.11 Joint simulation: simulated surplus paths; $n_{\text{sim}} = 50$

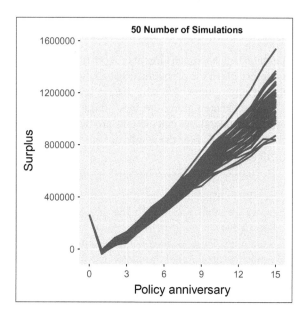

Fig. 6.12 Joint simulation: from simulated surplus paths to the empirical distribution of the surplus at maturity ($t = 15$); $n_{\text{sim}} = 50$

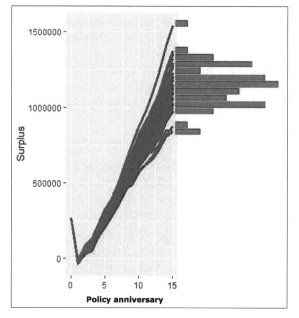

Fig. 6.13 Joint simulation: distribution of the surplus at time $t = 5$; $n_{\text{sim}} = 10\,000$

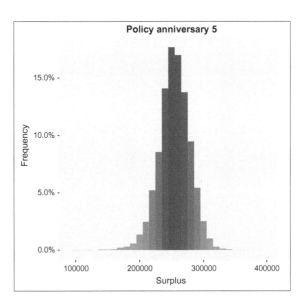

to the geometric Brownian motion model for the investment yield and the binomial distributions for the numbers of deaths and the numbers of lapses/surrenders, with parameters representing the best-estimate assumptions. The basic idea underlying the construction of the empirical distributions of the surplus is sketched in Fig. 6.12, where the surplus at maturity is referred to. The distributions of the surplus at times $t = 5$, $t = 10$ and $t = 15$, are represented in Figs. 6.13, 6.14 and 6.15, respectively. Looking at the ranges of simulated outcomes, it clearly appears that both the mean value and the dispersion of the distribution increase with time. This result also appears looking at the set of simulated paths in Fig. 6.11. ∎

Example 6.8 Some results of marginal simulations are sketched in Figs. 6.16, 6.17 and 6.18. Again, in each figure, n_{sim} denotes the number of simulations. Looking at the ranges of simulated outcomes, it clearly appears that the largest range pertains to the investment yield. In particular, comparing the distribution in Fig. 6.16 to the one in Fig. 6.15, we can realize that the investment yield constitutes the most significant contribution to the surplus randomness, and hence the most important cause of risk. As already noted, this is due to the structure of the insurance product, in which the financial component plays an important role. In Sect. 6.2, we will see, from another perspective, how an appropriate RM action, aiming to weaken the financial guarantee embedded in this product, can significantly reduce the randomness of results of interest. ∎

Remark Example 6.7 shows the application of the stochastic approach 2b (see Sect. 4.4.4) via a joint simulation procedure. Implementations of the approach 2a are shown in Example 6.8, via marginal simulations.

Fig. 6.14 Joint simulation: distribution of the surplus at time $t = 10$; $n_{sim} = 10\,000$

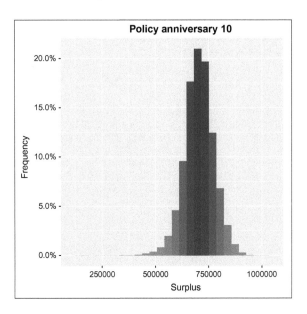

Fig. 6.15 Joint simulation: distribution of the surplus at time $t = 15$; $n_{sim} = 10\,000$

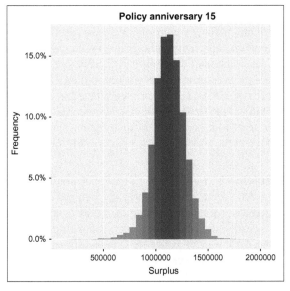

Fig. 6.16 Marginal
simulation of the investment
yield: distribution of the
surplus at time $t = 15$;
$n_{sim} = 10\,000$

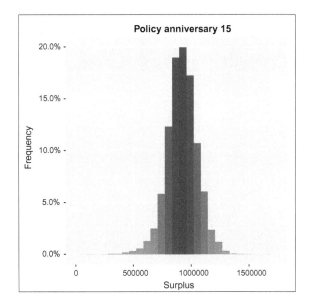

Fig. 6.17 Marginal
simulation of the mortality:
distribution of the surplus at
time $t = 15$; $n_{sim} = 10\,000$

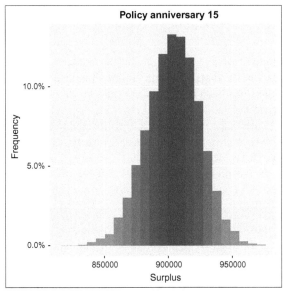

Fig. 6.18 Marginal
simulation of the
lapses/surrenders:
distribution of the surplus at
time $t = 15$; $n_{sim} = 10\,000$

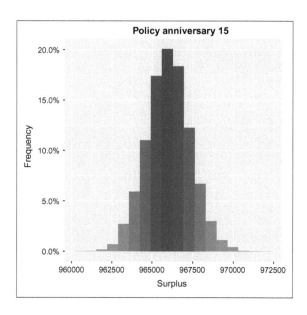

6.1.7 Introducing Systematic Deviations

Implementations of the stochastic model we have so far adopted only allow for
random fluctuations. Thus, no possible uncertainty, for example, in the parameter
values has been accounted for, and hence no possible systematic deviations have
been included in the model.

Systematic deviations can be modeled in a number of ways. Here a very simple
setting is chosen, which consists of a set of alternative scenarios, as discussed in
Sect. 6.1.4 (see Example 6.2).

In the presence of alternative scenarios, the stochastic approach described in
Sect. 6.1.6 can be applied separately to each scenario. The overall structure of the
simulation procedure can then be described in terms of two nested loops:

- the outer loop only aims to sequentially state the scenario;
- in the inner loop, all (or some of) the input variables are simulated according to
  the scenario stated in the outer loop.

The above structure is sketched in Fig. 6.19, where n_S is the number of scenarios,
each one simply denoted by $S_{(h)}$; the number of simulations conditional on each
scenario is denoted by n_{sim}.

This implementation yields three distinct distributions of the result of interest,
each distribution being conditional on a specific scenario. See Example 6.9.

Example 6.9 We consider three investment yield scenarios, namely I_1 (worst-
case), I_2 (best-estimate) and I_3 (best case); see Table 6.1 and assumptions (6.25).
Then, we simulate investment yields in the three scenarios, whereas mortality and

Fig. 6.19 Loop 1—
Implementation of a
stochastic approach, only
allowing for random
fluctuations

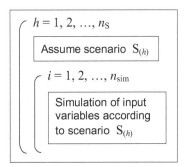

Fig. 6.20 Loop 2—
Implementation of a
double-stochastic approach,
allowing for systematic
deviations and random
fluctuations

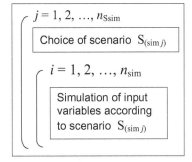

lapses/surrenders are simulated according to the best-estimate scenarios, i.e. D_2 and A_2 respectively. Some simulated paths of the surplus are shown in Fig. 6.21. The three conditional empirical distributions of the surplus at maturity are sketched in Fig. 6.22. ■

6.1.8   A Double-Stochastic Approach

A probability distribution can be assigned, according to expert's judgement, to the scenario space. Again, the overall structure of the simulation procedure can be described in terms of two nested loops:

- the outer loop aims at the random choice of the scenario, according to the probability distribution on the scenario space;
- in the inner loop, all (or some of) the input variables are simulated according to the scenario randomly chosen in the outer loop.

The above structure is sketched in Fig. 6.20, where n_{Ssim} is the number of scenario simulations; the scenario chosen in the j-th simulation is denoted by $S_{(sim j)}$; again, the number of simulations conditional on each scenario is denoted by n_{sim}.

All the results obtained while performing the inner loop are then automatically weighted with the frequencies of the scenarios under which they are achieved. Hence, unconditional distributions of the results of interest can be constructed. See Example 6.10.

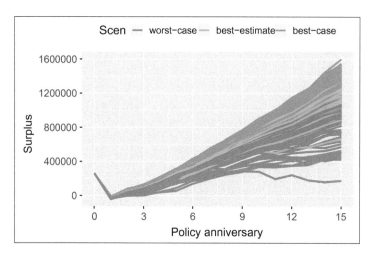

Fig. 6.21 Joint simulation of mortality, lapses/surrenders and investment yield in the three investment scenarios: simulated surplus paths; $n_{\text{sim}} = 100$

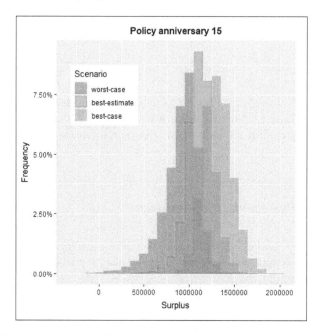

Fig. 6.22 Joint simulation of mortality, lapses/surrenders and investment yield in the three investment scenarios: distributions of the surplus at time $t = 15$; $n_{\text{sim}} = 10\,000$

Example 6.10 We consider three mortality scenarios, namely D_1 (best case), D_2 (best-estimate) and D_3 (worst case); see Table 6.2. The following probabilities are assigned to the three scenarios:

$$\mathbb{P}[D_1] = 0.20; \quad \mathbb{P}[D_2] = 0.60; \quad \mathbb{P}[D_3] = 0.20 \tag{6.36}$$

Fig. 6.23 Joint simulation, with double-stochastic approach for the mortality: unconditional distribution of surplus at time $t = 10$; $n_{Ssim} = 100$, $n_{sim} = 1\,000$

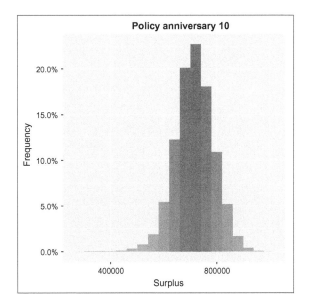

Fig. 6.24 Joint simulation, with double-stochastic approach for the mortality: unconditional distribution of surplus at time $t = 15$; $n_{Ssim} = 100$, $n_{sim} = 1\,000$

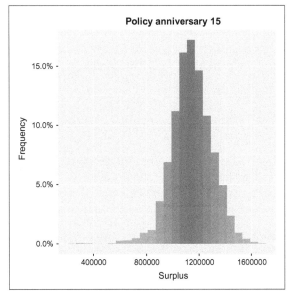

The unconditional empirical distributions of the surplus at time $t = 10$ and $t = 15$ are shown in Figs. 6.23 and 6.24, respectively. It is interesting to compare the distribution in Fig. 6.24 to the distribution in Fig. 6.17: the uncertainty about the mortality scenario implies a higher variability of the results. ∎

Fig. 6.25 Surplus paths in the best-estimate scenario (S_{222}), depending on two participation shares ($\eta = 0.70$, $\eta = 0.95$)

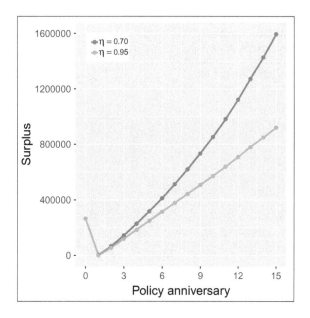

Remark Uncertainty in future scenarios and hence possible systematic deviations are accounted for in Example 6.9, by adopting approach 3 (see Sect. 4.4.4). Conversely, approach 4 is shown in Example 6.10.

6.1.9   Checking the Impact of Decision Variables

So far we have assumed fixed values (in particular given by the first-order technical basis) for all the decision variables involved in the various portfolio calculations. All the approaches we have presented and discussed in the previous sections also allow us to check the impact of decision variables on the output results of interest.

Example 6.11 We look at the impact of the participation share η in the portfolio surplus. In particular, we consider the reduction from $\eta = 95\%$ (so far assumed, see Example 6.1) to $\eta = 70\%$. Of course, the lower policyholders' participation in the financial profits implies a higher surplus accumulation for the insurer, as it appears from Fig. 6.25 (deterministic approach) and Fig. 6.26 (stochastic approach with marginal simulation of the investment yield according to the best-estimate scenario). ∎

Fig. 6.26 Marginal simulation of the investment yield: distribution of the surplus at $t = 15$, depending on two participation shares ($\eta = 0.70$, $\eta = 0.95$); $n_{\text{sim}} = 10\,000$

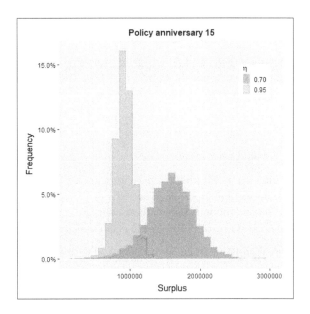

6.2   A Case Study: Assessing the Impact of Alternative Interest Guarantees

The problem we are addressing provides an interesting example of implementation of both the deterministic approach and the stochastic approach to the choice of an appropriate product design. The problem has been dealt with by Brighenti (2015).

6.2.1   Basic Ideas

We consider a portfolio of endowment insurance policies. A profit participation mechanism works. The participation is only based on the insurer's investment yield provided by a segregated fund.

The profit participation is implemented via an increment in the policy reserve (also known as "reserve revaluation"), which in its turn determines an increment in the benefits. See Sect. 6.1.1 for details. We assume constant annual premiums payable for the whole policy duration.

Let i denote the technical rate of interest, adopted in premium and reserve calculations. The following quantities are defined for $t = 1, 2, \ldots, m$, where m denotes the policy maturity:

g_t = investment yield in year $(t - 1, t)$;
η_t = participating share;

r_t = increase in the policy reserve (beyond the technical interest rate i), also known as the "revaluation rate".

According to the traditional participation formula, the annual rate of increase of the mathematical reserve is given by Eq. (6.6). Then, as seen in Sect. 6.1.1, the revaluation rate is given by:

$$r_t = \max\left\{\frac{\eta_t\, g_t - i}{1 + i}, 0\right\} \tag{6.37}$$

We note what follows:

- Equation (6.37) implies that the interest i is annually guaranteed, and hence the extra-yield $\eta_t\, g_t - i$ is locked-in;
- a cliquet option is then embedded in the endowment insurance product;
- usually, the cost of the option is not explicitly charged to policyholders.

A slightly modified version of (6.37) is the following one:

$$r_t = \max\left\{\frac{\eta_t\, g_t - i}{1 + i}, r_{\min}\right\} \tag{6.38}$$

where $r_{\min}$ denotes the minimum annual revaluation rate.

A possible approach aiming to mitigate the market risk borne by the insurer because of the lock-in mechanism, consists in weakening the interest rate guarantee.

6.2.2   Alternative Interest Rate Guarantees

The mathematical reserve of a non-participating insurance policy is the result of the financial accumulation of the savings premiums $P_s^{[S]}$, according to the following relation:

$$V_t = \sum_{s=0}^{t-1} P_s^{[S]} (1 + i)^{t-s} \tag{6.39}$$

According to a more general setting, the result of the accumulation process can be defined as follows:

$$V_t = \sum_{s=0}^{t-1} P_s^{[S]} f(s, t) \tag{6.40}$$

where $f(s, t)$ denotes a generic accumulation factor over the time interval (s, t).

A number of interest rate guarantees can be defined in terms of accumulation factors. We only focus on the benefit in case of survival at maturity. We assume a constant participation share, that is, $\eta_t = \eta$ for $t = 1, 2, \ldots, m$.

We define the following accumulation factors.

- No profit participation:

$$f^{[0]}(s, t) = (1 + i)^{t-s} \tag{6.41}$$

- Traditional interest rate guarantee, according to (6.6) and (6.37):

$$f^{[1]}(s, t) = \prod_{h=s+1}^{t} (1 + \max\{\eta\, g_h, i\}) = \prod_{h=s+1}^{t} \max\{(1 + i), (1 + \eta\, g_h)\} \tag{6.42}$$

- Traditional interest rate guarantee with minimum revaluation rate $r_{\min}$:

$$f^{[2]}(s, t) = \prod_{h=s+1}^{t} \max\{(1 + i)\,(1 + r_{\min}), (1 + \eta\, g_h)\} \tag{6.43}$$

- No interest guarantee:

$$f^{[3]}(s, t) = \prod_{h=s+1}^{t} (1 + \eta\, g_h) \tag{6.44}$$

- Maturity guarantee (at time m):

$$f^{[4]}(s, t) = \begin{cases} \prod_{h=s+1}^{t}(1 + \eta\, g_h) & \text{if } s < t < m \\ \max\left\{\prod_{h=s+1}^{m}(1 + \eta\, g_h), (1 + i)^{m-s}\right\} & \text{if } t = m \end{cases} \tag{6.45}$$

We note that the accumulation factor defined by (6.44) implies absence of guarantees, and hence can only be used to compare the effect of the various accumulation factors, and to define a specific accumulation process prior to maturity (see Eq. (6.45)).

The following inequalities hold:

$$f^{[1]}(s, t) \geq f^{[0]}(s, t) \tag{6.46a}$$

$$f^{[2]}(s, t) \geq f^{[0]}(s, t) \tag{6.46b}$$

$$f^{[3]}(s, t) \gtrless f^{[0]}(s, t) \tag{6.46c}$$

$$f^{[3]}(s, t) \leq f^{[1]}(s, t) \tag{6.46d}$$

$$f^{[3]}(s, t) \leq f^{[2]}(s, t) \tag{6.46e}$$

$$f^{[4]}(s, m) \geq (1 + i)^{m-s} \tag{6.46f}$$

Profit participation outcomes of course depend on the investment yield scenario, as we will see in the numerical examples in the next sections.

6.2.3   Deterministic Approach

A first insight into the effect of the various accumulation factors can be provided by adopting a deterministic approach, that is, via assumptions on the investment yield scenario, that is, the sequence of yields g_t, $t = 1, 2, \ldots, m$.

Example 6.12 Three alternative assumptions underpin the results, in terms of the accumulation factors $f^{[\cdot]}(0, t)$, shown in Tables 6.4, 6.5 and 6.6, respectively. The participation share is $\eta = 0.95$. The paths of the accumulation factors $f^{[0]}(0, t)$, $f^{[1]}(0, t)$ and $f^{[4]}(0, t)$ are plotted in Figs. 6.27, 6.28 and 6.29, respectively. The diversity of the impact of the various accumulation factors, and hence of the various guarantee mechanisms, is self-evident. In particular, the impact of the lock-in mechanism clearly appears in Figs. 6.28 and 6.29. ∎

Remark Example 6.12 shows the results of a what-if analysis (see Sect. 4.4.6) via implementation of the deterministic approach 1a (see Sect. 4.4.4). In each implementation, three "values" are assigned to a decision variable (see the scheme in Fig. 4.16), each "value" being given by a specific participation mechanism.

Table 6.4 Accumulation factors in a high-yield scenario

t	g_t	$f^{[0]}(0, t)$ $i = 0.02$	$f^{[1]}(0, t)$ $i = 0.02$	$f^{[2]}(0, t)$ $i = 0$ $r_{min} = 0.02$	$f^{[3]}(0, t)$	$f^{[4]}(0, t)$ $i = 0.02$
1	5.00%	1.0200	1.0475	1.0475	1.0475	1.0475
2	4.50%	1.0404	1.0923	1.0923	1.0923	1.0923
3	4.00%	1.0612	1.1338	1.1338	1.1338	1.1338
4	3.00%	1.0824	1.1661	1.1661	1.1661	1.1661
5	2.80%	1.1041	1.1971	1.1971	1.1971	1.1971
6	2.90%	1.1262	1.2301	1.2301	1.2301	1.2301
7	3.50%	1.1487	1.2710	1.2710	1.2710	1.2710
8	2.70%	1.1717	1.3036	1.3036	1.3036	1.3036
9	3.10%	1.1951	1.3420	1.3420	1.3420	1.3420
10	3.30%	1.2190	1.3841	1.3841	1.3841	1.3841
11	2.50%	1.2434	1.4169	1.4169	1.4169	1.4169
12	2.50%	1.2682	1.4506	1.4506	1.4506	1.4506
13	2.90%	1.2936	1.4906	1.4906	1.4906	1.4906
14	3.40%	1.3195	1.5387	1.5387	1.5387	1.5387
15	4.50%	1.3459	1.6045	1.6045	1.6045	1.6045

Table 6.5 Accumulation factors in a medium-yield scenario

t	g_t	$f^{[0]}(0, t)$ $i = 0.02$	$f^{[1]}(0, t)$ $i = 0.02$	$f^{[2]}(0, t)$ $i = 0$ $r_{min} = 0.02$	$f^{[3]}(0, t)$	$f^{[4]}(0, t)$ $i = 0.02$
1	2.50%	1.0200	1.0238	1.0238	1.0238	1.0238
2	2.50%	1.0404	1.0481	1.0481	1.0481	1.0481
3	3.00%	1.0612	1.0779	1.0779	1.0779	1.0779
4	2.70%	1.0824	1.1056	1.1056	1.1056	1.1056
5	1.80%	1.1041	1.1277	1.1277	1.1245	1.1245
6	1.30%	1.1262	1.1502	1.1502	1.1384	1.1384
7	1.20%	1.1487	1.1733	1.1733	1.1514	1.1514
8	1.00%	1.1717	1.1967	1.1967	1.1623	1.1623
9	1.10%	1.1951	1.2207	1.2207	1.1744	1.1744
10	1.50%	1.2190	1.2451	1.2451	1.1912	1.1912
11	2.30%	1.2434	1.2723	1.2723	1.2172	1.2172
12	3.40%	1.2682	1.3134	1.3134	1.2565	1.2565
13	2.40%	1.2936	1.3433	1.3433	1.2852	1.2852
14	3.50%	1.3195	1.3880	1.3880	1.3279	1.3279
15	4.50%	1.3459	1.4473	1.4473	1.3847	1.3847

Table 6.6 Accumulation factors in a low-yield scenario

t	g_t	$f^{[0]}(0, t)$ $i = 0.02$	$f^{[1]}(0, t)$ $i = 0.02$	$f^{[2]}(0, t)$ $i = 0$ $r_{min} = 0.02$	$f^{[3]}(0, t)$	$f^{[4]}(0, t)$ $i = 0.02$
1	2.20%	1.0200	1.0209	1.0209	1.0209	1.0209
2	2.30%	1.0404	1.0432	1.0432	1.0432	1.0432
3	2.00%	1.0612	1.0641	1.0641	1.0630	1.0630
4	3.10%	1.0824	1.0954	1.0954	1.0943	1.0943
5	2.50%	1.1041	1.1214	1.1214	1.1203	1.1203
6	1.80%	1.1262	1.1439	1.1439	1.1395	1.1395
7	1.20%	1.1487	1.1667	1.1667	1.1525	1.1525
8	1.30%	1.1717	1.1901	1.1901	1.1667	1.1667
9	1.50%	1.1951	1.2139	1.2139	1.1833	1.1833
10	1.00%	1.2190	1.2381	1.2381	1.1946	1.1946
11	1.50%	1.2434	1.2629	1.2629	1.2116	1.2116
12	1.20%	1.2682	1.2882	1.2882	1.2254	1.2254
13	1.50%	1.2936	1.3139	1.3139	1.2429	1.2429
14	1.00%	1.3195	1.3402	1.3402	1.2547	1.2547
15	1.20%	1.3459	1.3670	1.3670	1.2690	1.3459

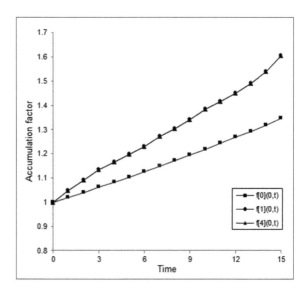

Fig. 6.27 High-yield scenario

6.2.4 Stochastic Approach

The deterministic approach does not provide any information about the portfolio risk profile. A stochastic approach is needed in order to explore the impact of diverse interest guarantees on the risk profile.

First, a quantity must be chosen to summarize the portfolio results. We will address the PVFP, that is the present value of future profits originated by the portfolio. Given the random nature of several input quantities, in particular the investment yields, the PVFP is a random variable. We aim at finding the probability distribution of the PVFP.

We assume:

- deterministic mortality;
- no surrenders.

Further, we disregard expenses and related loadings.

A stochastic simulation procedure must be defined to generate sequences of investment yields, and then to find the empirical distribution of the PVFP. To this purpose, we generalize the setting proposed in Sect. 6.1.4.

The investment yields depend on the mix of assets which constitute the segregated fund. We assume that the mix is given by:

$$(\text{risk-free bonds, other bonds, equities})$$

and the relevant shares are:

$$\left(\alpha, \ \beta, \ \gamma = 1 - (\alpha + \beta)\right) \tag{6.47}$$

Fig. 6.28 Medium-yield scenario

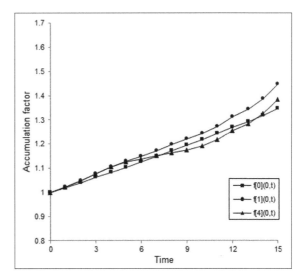

Fig. 6.29 Low-yield scenario

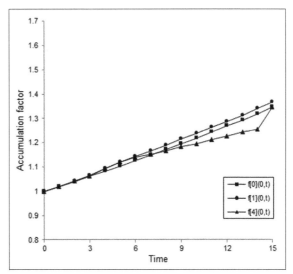

We assume that the shares are constant throughout the portfolio duration.

The values at time t (and hence the performances up to time t) of the three types of assets are represented by the following quantities:

- $M(t)$ for risk-free bonds;
- $B(t)$ for other bonds;
- $S(t)$ for equities.

For the investment mix we then have:

$$I(t) = \alpha\, M(t) + \beta\, B(t) + \gamma\, S(t) \tag{6.48}$$

The random yield of the segregated fund is defined as follows:

$$g_t = \frac{I(t+1)}{I(t)} - 1 \tag{6.49}$$

We assume a deterministic behavior for $M(t)$:

$$M(t) = M(0)\, e^{\delta t} \tag{6.50}$$

where δ denotes the (constant) force of interest.

A geometric Brownian motion is conversely assumed to describe the behavior of $B(t)$ and $S(t)$:

$$dB(t) = \mu_B\, B(t)\, dt + \sigma_B\, B(t)\, dW_B(t) \tag{6.51}$$
$$dS(t) = \mu_S\, S(t)\, dt + \sigma_S\, S(t)\, dW_S(t) \tag{6.52}$$

where $W_B(t)$ and $W_S(t)$ are correlated Wiener processes:

$$dW_B(t), dW_S(t) \sim N(0, dt) \tag{6.53}$$
$$\mathrm{Cov}\big(dW_B(t), dW_S(t)\big) = \rho\, dt \tag{6.54}$$

We note that:

μ_B, μ_S are the drift parameters, representing the estimated yields;
σ_B, σ_S are the diffusion parameters, representing the volatility;
ρ is the correlation coefficient.

We assume, as reasonable:

$$\mu_B < \mu_S \tag{6.55}$$
$$\sigma_B < \sigma_S \tag{6.56}$$

Example 6.13 We refer to a portfolio of $N_0 = 1\,000$ endowment insurance policies, with initial sum assured $C = 1$ and maturity $m = 15$. All the policyholders are age 50 at policy issue. We assume (in all the numerical examples):

$\delta = 0.005;$
$\mu_B = 0.025; \quad \sigma_B = 0.0105;$
$\mu_S = 0.040; \quad \sigma_S = 0.0550.$

Technical rate of interest i and minimum revaluation rate r_{min} can alternatively take the values displayed in Table 6.7.

The age-pattern of mortality follows the first Heligman–Pollard law (see Eq. 4.9). The parameters of the first-order mortality, adopted in premium and reserve calculations, are specified in Table 4.1; some relevant markers are displayed in Table 4.2.

Table 6.7 Interest rate and minimum revaluation rate

i	r_{min}
0.02	0
0	0.02

Table 6.8 Parameters of the first Heligman–Pollard law: best-estimate mortality

A	B	C	D	E	F	G	H
0.00054	0.01700	0.10100	0.00013	10.72	18.67	1.464 E−05	1.11000

Table 6.9 Some markers of the first Heligman–Pollard law: best-estimate mortality

$\overset{\circ}{e}_0$	$\overset{\circ}{e}_{40}$	$\overset{\circ}{e}_{65}$	Mode	q_0	q_{40}	q_{80}
79.412	40.653	18.352	85	0.00684	0.00097	0.05826

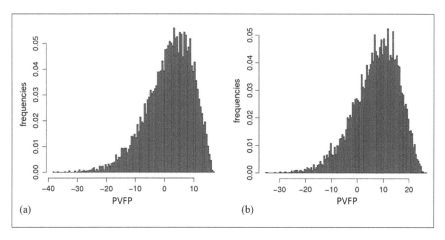

Fig. 6.30 Assets = (10, 80, 10); participation mechanism $f^{[1]}$; $i = 0.02$ **a** $\eta = 0.95$, **b** $\eta = 0.90$

The parameters which describe the best-estimate mortality are shown in Table 6.8; some relevant markers are displayed in Table 6.9.

In Figs. 6.30, 6.31, 6.32, 6.33, 6.35, 6.36, 6.37 and 6.38 the (empirical) distributions of the PVFP are plotted. Various combinations are assumed, regarding:

- the asset mix (α, β, γ);
- the participation mechanism $f^{[\cdot]}$ (and hence the interest guarantee);
- the interest rate i and the minimum revaluation rate r_{min};
- the profit participation share η (assumed constant over the whole policy duration).

The impact of diverse combinations is self-evident. In particular, we note what follows.

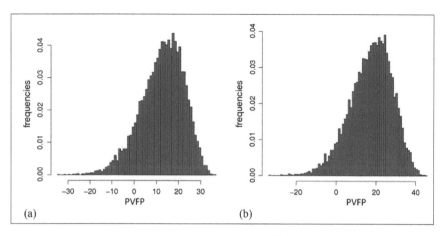

Fig. 6.31 Assets = (10, 80, 10); participation mechanism $f^{[1]}$; $i = 0.02$ **a** $\eta = 0.85$, **b** $\eta = 0.80$

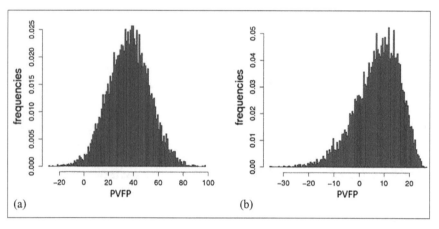

Fig. 6.32 Assets = (10, 80, 10); participation share $\eta = 0.90$ **a** $f^{[0]}$; $i = 0.02$, **b** $f^{[1]}$; $i = 0.02$

- A reduction in the participation share η implies a general increase in the PVFP, as clearly appears looking at Figs. 6.30a, b, 6.31a, b.
- The absence of profit participation of course increases the PVFP and, in particular, its variability: see Fig. 6.32a, b.
- Different combinations of i and r_{min} (with $i + r_{min}$ = const.) have no significant impact on the PVFP, as appears by comparing Fig. 6.35a, b, or Fig. 6.37a, b.
- Changing the mix of assets impacts on the PVFP distribution; in particular a higher equity share increases the expectation as well as the variability; see, for example, Figs. 6.37a and 6.35a.

Fig. 6.33 Assets = (10, 80, 10); participation share $\eta = 0.90$; $f^{[2]}$; $i = 0$; $r_{\min} = 0.02$

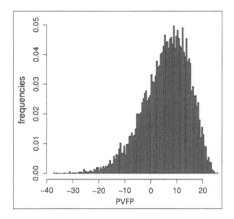

Fig. 6.34 Assets = (10, 80, 10); participation share $\eta = 0.90$; $f^{[4]}$; $i = 0.02$

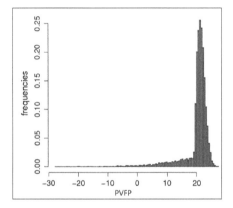

- Guarantee weakening by adopting accumulation factor $f^{[4]}$ has a significant impact on the PVFP distribution, and specifically on its variability, as clearly appears comparing, for example, the result Fig. 6.34 to the ones in Figs. 6.32a, b and 6.33, or the result in Fig. 6.36 to the ones in Fig. 6.35a, b, or the result in Fig. 6.38 to the ones in Fig. 6.37a, b.

■

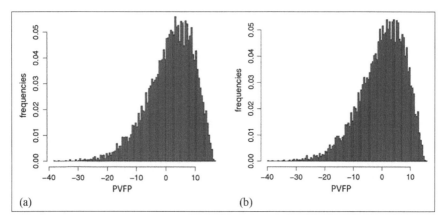

Fig. 6.35 Assets = $(10, 80, 10)$;   participation share $\eta = 0.95$ **a** $f^{[1]}$; $i = 0.02$, **b** $f^{[2]}$; $i = 0$; $r_{min} = 0.02$

Fig. 6.36 Assets = $(10, 80, 10)$; participation share $\eta = 0.95$; $f^{[4]}$; $i = 0.02$

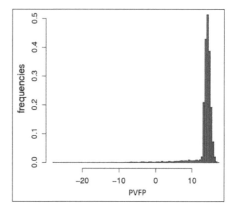

Remark A stochastic model, in line with the approach 2a (see Sect. 4.4.4), is used to obtain the PVFP empirical distributions in Example 6.13: marginal simulations are performed to sample the outcomes of the investment yield and the related impact on the PVFP for several participation mechanisms (that is, decision variables), while mortality is assumed deterministic.

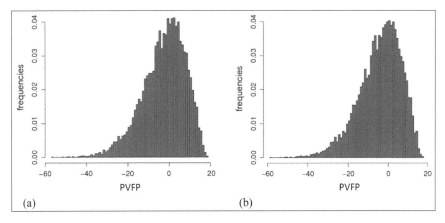

Fig. 6.37 Assets = $(10, 70, 20)$; participation share $\eta = 0.95$ **a** $f^{[1]}$; $i = 0.02$, **b** $f^{[2]}$; $i = 0$; $r_{\min} = 0.02$

Fig. 6.38 Assets = $(10, 70, 20)$; participation share $\eta = 0.95$; $f^{[4]}$; $i = 0.02$

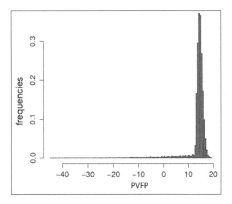

6.3   References and Suggestions for Further Reading

References on endowment insurance products have been provided in Sect. 5.10. We only recall recent papers focusing on participation mechanisms and interest rate guarantees: Reuss et al. (2015, 2016) (and the extended lists of references therein), Gatzert et al. (2012a), Heling and Holder (2013), and Maurer et al. (2013).

The stochastic model which assesses the impact of diverse participation mechanisms on the PVFP of an endowment portfolio was developed and implemented by Brighenti (2015).

Chapter 7
Risk Assessment and Impact Assessment in Life Annuity Business

7.1 A Stochastic Model for a Life Annuity Portfolio

Following the structure adopted in Sect. 6.1, we define all the quantities needed to construct a specific model for a life annuity portfolio, in the framework proposed in Sect. 4.4.3. We then illustrate various implementations of the model, aiming at impact assessment.

7.1.1 The Portfolio

We refer to a portfolio of single-premium immediate life annuities, consisting of one generation. All the policies are issued at time 0; x_0 denotes the age at policy issue.

According to policy conditions, the following quantities are defined:

b = annual benefit, payable at the end of each year (i.e., at times $t = 1, 2, \ldots$);
Π = net single premium;
V_t = policy reserve at time t ($t = 0, 1, \ldots$).

As we aim at focusing on the impact of the longevity risk, we disregard expenses and related loadings, as well as profit participation mechanisms.

The premium and the policy reserve are calculated, according to the equivalence principle, as follows (the traditional actuarial notation is adopted).

- Net single premium:

$$\Pi = b\, a_{x_0} = b \sum_{h=1}^{+\infty} (1+i)^{-h}\, {}_h p_{x_0} \tag{7.1}$$

All the examples in this chapter have been planned and elaborated by Daniela Y. Tabakova.

© Springer Nature Switzerland AG 2020
E. Pitacco, *ERM and QRM in Life Insurance*, Springer Actuarial,
https://doi.org/10.1007/978-3-030-49852-8_7

where i denotes the technical rate of interest, and $_h p_{x_0}$ is the probability of a person age x_0 at policy issue being alive at age $x_0 + h$.

- Policy reserve at time t:

$$V_t = b \, a_{x_0+t} \qquad (7.2)$$

The following random quantities are defined at portfolio level:

N_t = number of policies in force at time t, with initial number N_0 given;
D_t = number of deaths in year t, i.e., between $t - 1$ and t.

Then:

$$N_t = N_{t-1} - D_t \qquad (7.3)$$

We note that surrenders are not allowed in single-premium immediate life annuities.
   At portfolio level we have:

$\Pi^{[P]} = \Pi \, N_0$ = total amount of premiums at time 0;

and, for $t = 1, 2, \ldots$:

$B_t^{[P]} = b \, N_t$ = total amount of annuity benefits paid at time t;
$V_t^{[P]} = V_t \, N_t$ = portfolio reserve at time t (with $V_0^{[P]} = \Pi_0^{[P]}$).

Moreover, we define the following random quantities:

$F_t^{[P]}$ = portfolio fund at time t; see Eqs. (7.5) and (7.6);
M_t = surplus, that is, shareholders' capital (assigned to the portfolio) at time t:

$$M_t = F_t^{[P]} - V_t^{[P]} \qquad (7.4)$$

We assume absence of shareholders' capital flows (allocations or releases), besides (possible) initial allocation, which leads to:

$$F_0^{[P]} = \Pi^{[P]} + M_0 \qquad (7.5)$$

Remark 7.1 As regards the definition of the portfolio reserve, $V_t^{[P]} = V_t \, N_t$, the reader is referred to Remark 6.1 in Sect. 6.1.1.

The portfolio fund dynamics, for $t = 1, 2, \ldots$, is defined as follows:

$$F_t^{[P]} = F_{t-1}^{[P]} (1 + I_t) - B_t^{[P]} \qquad (7.6)$$

where I_t denotes the random investment yield. From recursion (7.6) we obtain:

$$F_t^{[P]} = (\Pi^{[P]} + M_0) \prod_{h=1}^{t} (1 + I_h) - \sum_{h=1}^{t} \left[B_h^{[P]} \prod_{j=h}^{t} (1 + I_j) \right] \qquad (7.7)$$

Remark 7.2 Equation (7.7) provides another example of aggregation function (see Sect. 4.4.3); in particular:

- $F_t^{[P]}$ is the (annual) result of interest;
- Π (which determines $\Pi^{[P]}$) and M_0 are decision variables;
- the benefits $B_h^{[P]}$ depend on the number of survivors, i.e., a scenario variable;
- the random yields I_h are scenario variables.

7.1.2   From Theory to Practice

The examples in the following sections aim at showing simple applications of QRM to a life annuity portfolio, in line with the approaches described in Sect. 4.4. In particular:

- The structure of a specific annuity portfolio is defined in Example 7.1; technical bases, premiums and policy reserves are provided.
- Various longevity scenarios are defined in Example 7.2.
- Following the approaches described in Sect. 4.4.4, we present:

  - results of a deterministic analysis in Examples 7.3, 7.4 and 7.5;
  - results of a stochastic analysis, only allowing for random fluctuations, in Example 7.6;
  - results obtained by allowing for uncertainty, and hence systematic deviations, in Example 7.7;
  - results achieved by implementing a double-stochastic approach in Examples 7.8 and 7.9.

- An analysis of the impact of the technical rate of interest as a decision variable, facing diverse longevity scenarios, is presented in Example 7.10.
- Finally, Examples 7.11 and 7.12 focus on solvency issues and capital requirements.

7.1.3   Defining an Annuity Portfolio

In the following example we define the annuity portfolio which will be referred to in the risk assessment and impact assessment examples.

Example 7.1 We refer to the generic portfolio described in Sect. 7.1.1, and assume the following data:

- Initial portfolio size: $N_0 = 1\,000$;
- Annual benefit amount: $b = 100$;
- Initial age: $x_0 = 70$;
- Technical basis adopted in premium and reserve calculations:

Table 7.1 Parameters of the third term of the first Heligman–Pollard law: first-order technical basis

G	H
1.06038 E−06	1.13705

Table 7.2 Some markers of the first Heligman–Pollard law: first-order technical basis

$\overset{\circ}{e}_0$	$\overset{\circ}{e}_{65}$	Mode	q_{80}
86.464	23.389	91	0.02984

▷ interest rate: $i = 0.02$;

▷ mortality assumption: third term of the first Heligman–Pollard law, that is

$$\frac{q_x}{1 - q_x} = G\,H^x \qquad (7.8)$$

with the parameter values specified in Table 7.1; the related life table is a projected life table, and the resulting modal age at death is approximately 91 (see Table 7.2).

The single premium is then $\Pi = 1490.71$. ∎

7.1.4 Scenarios

As already mentioned, we only focus on the impact of longevity risk (in terms of both random fluctuations and systematic deviations). Hence, we assume a deterministic investment yield, instead of the random yield denoted by I_t in Eq. (7.6). It follows that all random results only depend on mortality in the portfolio.

Example 7.2 The investment yield is assumed equal to the technical rate of interest, that is 0.02. As regards annuitants' mortality, we assume as the best-estimate scenario the age-pattern of mortality q_x^* given by the third term of the Heligman–Pollard law with the parameter values specified in Table 7.3; the related life table is a projected life table, and the resulting modal age at death is approximately 90 (see Table 7.4).

Table 7.3 Parameters of the third term of the first Heligman–Pollard law: best-estimate mortality scenario

G	H
2.00532 E−06	1.13025

Table 7.4 Some markers of the first Heligman–Pollard law: best-estimate mortality scenario

$\overset{\circ}{e}_0$	$\overset{\circ}{e}_{65}$	Mode	q_{80}^*
85.128	22.350	90	0.03475

We note that the mortality component of the first-order basis, used in premium and reserve calculation, represents a prudential choice with respect to the best-estimate mortality. Alternative mortality scenarios, considered in the following examples, are as follows:

$$S_{(1)} = \{0.75\,q_x^*\}, \ S_{(2)} = \{0.90\,q_x^*\}, \ S_{(3)} = \{q_x^*\}, \ S_{(4)} = \{1.10\,q_x^*\}, \ S_{(5)} = \{1.25\,q_x^*\} \tag{7.9}$$

Each scenario consists of a set of annual probabilities of death, for $x = 70, 71, \ldots$. Scenario $S_{(3)}$, defined by the parameters specified in Table 7.3, is assumed as the best-estimate scenario.

The portfolio reserves, based on the first-order assumption ("prudential" reserve) and the best-estimate assumption ("realistic" reserve) respectively, are plotted in Fig. 7.1, for $70 \leq x \leq 85$. Both the portfolio reserves are calculated assuming expected numbers of survivors determined according to the best-estimate annuitants' mortality. ∎

Fig. 7.1 Portfolio reserves

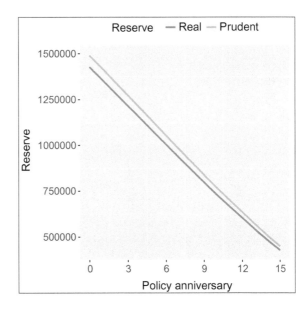

7.1.5 Deterministic Approach

The simplest implementation of the deterministic approach consists in a single mortality assumption. Moreover, the impacts of various mortality scenarios can be assessed and compared.

Example 7.3 The path of the surplus M_t, with $M_0 = 0$, originated by the best-estimate scenario $S_{(3)} = \{q_x^*\}$, is sketched in Fig. 7.2. Three different scenarios conversely underpin the paths in Fig. 7.3. Of course, a mortality lower than the best-estimate one implies a worse surplus path. ∎

Example 7.4 Expected annual cash flows, $\mathbb{E}[B_t]$, are compared in Fig. 7.4. The blue line represents the expected cash flows according to the first-order technical basis, while the histogram represents the expected cash flows according to the best-estimate mortality scenario: the prudential character of the first-order basis clearly emerges. The impact of a stress scenario on the annual cash flows is sketched in Fig. 7.5, where a decrease of 25% in the probabilities of death is assumed to occur from time $t = 10$, i.e., from age 80 onwards, that is, $q_x^* \rightarrow 0.75\, q_x^*$ for $x = 80, 81, \ldots$ ∎

Example 7.5 Shareholders capital allocation provides additional resources to meet an unexpected decrease in mortality. An appropriate capital allocation determines a raise of 6 000 monetary units in the threshold which represents a maintainable level of benefit payments (see Sect. 4.6, where the action is represented by path $(2) \rightarrow (a)$), as shown in Fig. 7.6. ∎

Fig. 7.2 Surplus path in the best-estimate scenario

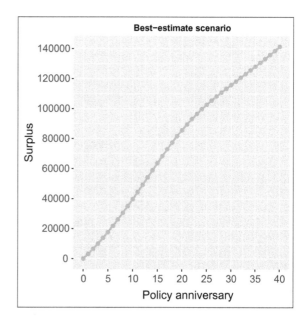

Fig. 7.3 Surplus paths in
three different scenarios

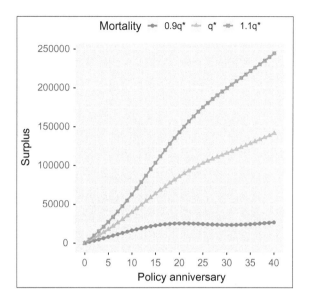

Fig. 7.4 Expected cash
flows according to the
first-order basis (line) versus
expected cash flows in
best-estimate scenario (bars)

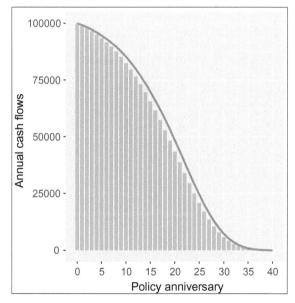

Fig. 7.5 Expected cash
flows according to the
first-order basis (line) versus
expected cash flows in a
stress scenario (bars)

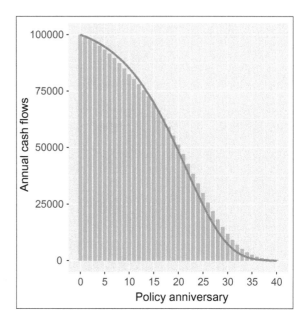

Fig. 7.6 Meeting expected
cash flows in a stress
scenario

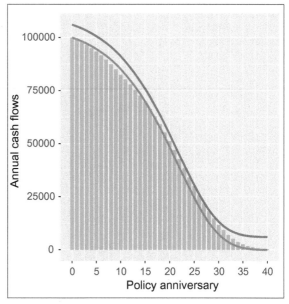

Remark Deterministic approaches 1a and 1b (see Sect. 4.4.4) are implemented, for each policy year, in Example 7.3, while Examples 7.4 and 7.5 show the impact of a stress scenario, still in line with approach 1a.

7.1.6 Stochastic Approach

As noted in Sect. 6.1.6, the stochastic approach aims at the construction of probability distributions of results of interest, given the probability distribution of the input variables. In our simplified setting, randomness is only due to the annual numbers of deaths, which are simulated according to the binomial distribution with parameter values provided by the assumed mortality scenario. Thus, the impact of random fluctuations in mortality is captured.

Example 7.6 Simulated surplus paths, according to the best-estimate mortality, are sketched in Figs. 7.7 and 7.8, where the results of 10 and 100 simulations are respectively shown. Empirical distribution of the surplus at various times are provided by Figs. 7.9, 7.10 and 7.11. It is interesting to note the increase in the variance, when moving from $t = 10$ to $t = 30$. ∎

Fig. 7.7 Mortality simulation: simulated surplus paths; $n_{\text{sim}} = 10$

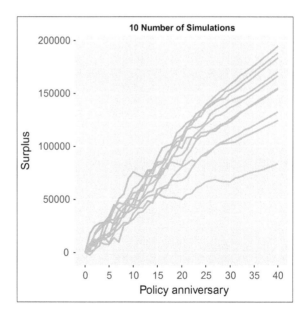

Fig. 7.8 Mortality
simulation: simulated
surplus paths; $n_{sim} = 100$

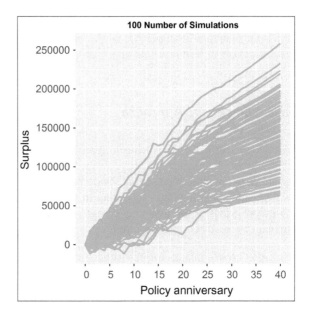

Fig. 7.9 Mortality
simulation: distribution of
the surplus at time $t = 10$;
$n_{sim} = 10\,000$

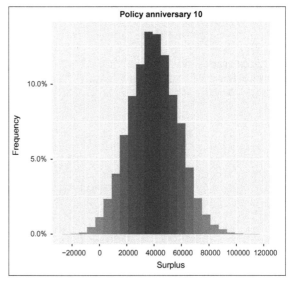

Fig. 7.10 Mortality
simulation: distribution of
the surplus at time $t = 20$;
$n_{sim} = 10\,000$

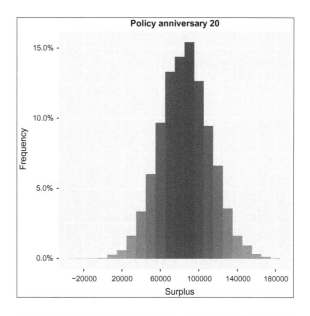

Fig. 7.11 Mortality
simulation: distribution of
the surplus at time $t = 30$;
$n_{sim} = 10\,000$

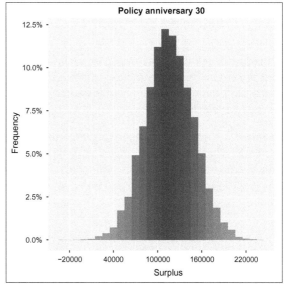

Fig. 7.12 Mortality
simulation in three different
mortality scenarios:
simulated surplus paths;
$n_{\text{sim}} = 3 \times 10$

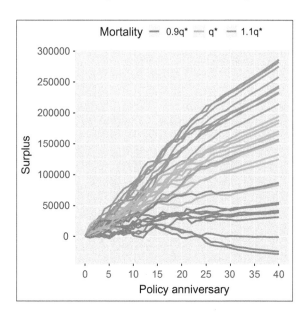

7.1.7   Introducing Systematic Deviations

In the framework of longevity impact analysis, uncertainty in future mortality constitutes a critical issue. A simple way of representing uncertainty in future mortality consists in assuming various possible mortality scenarios, each scenario being quantified by a projected life table.

Example 7.7 Uncertainty about future mortality is represented in Fig. 7.12 by three scenarios, which result in three sets of simulated paths of the surplus (each set of paths representing random fluctuations only). This approach can be implemented by adopting the scheme shown in Fig. 6.19. ∎

7.1.8   A Double-Stochastic Approach

To obtain a probability distribution of the surplus also accounting for uncertainty in future mortality and hence systematic deviations, a probability distribution must be assigned to the scenario space. The logical scheme sketched in Fig. 6.20 must then be implemented.

Example 7.8 We assume the five mortality scenarios $S_{(h)}, h = 1, \ldots, 5$, defined by (7.9). The following probability distribution has been assigned:

Fig. 7.13 Mortality
simulation: distribution of
the surplus at time $t = 10$ via
double stochastic approach;
$n_{Ssim} = 1\,000$, $n_{sim} = 1\,000$

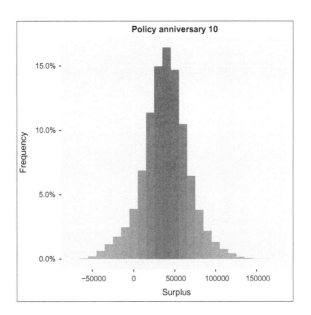

$$\mathbb{P}[S_{(1)}] = 0.05; \quad \mathbb{P}[S_{(2)}] = 0.15; \quad \mathbb{P}[S_{(3)}] = 0.60; \quad \mathbb{P}[S_{(4)}] = 0.15; \quad \mathbb{P}[S_{(5)}] = 0.05$$
$$(7.10)$$

Empirical distributions of the surplus at times $t = 10$ and $t = 15$ are illustrated in
Figs. 7.13 and 7.14.   ∎

Remark Stochastic approaches are implemented in Examples 7.6, 7.7, and 7.8. In
all the examples, only stochastic mortality is considered (while investment yield is
deterministic), hence marginal simulations are performed. In particular, only random fluctuations are accounted for in Example 7.6, according to approach 2a (see
Sect. 4.4.4), whereas systematic deviations are also allowed for in Examples 7.7 and
7.8, according to approaches 3 and 4, respectively.

Example 7.9 It is interesting to compare the empirical distribution which only
accounts for random fluctuations to the one which represents the impact of both
random fluctuations and systematic deviations. In Fig. 7.15, which refers to the surplus at time $t = 10$, the higher dispersion and, in particular, the higher probability
of negative results due to systematic deviations are evident.   ∎

7.1.9   Checking the Impact of Decision Variables

A number of tests can be performed to check the impact of decision variables on the
portfolio results. We only focus on a specific problem.

Fig. 7.14 Mortality
simulation: distribution of
the surplus at time $t = 15$ via
double stochastic approach;
$n_{Ssim} = 1\,000, n_{sim} = 1\,000$

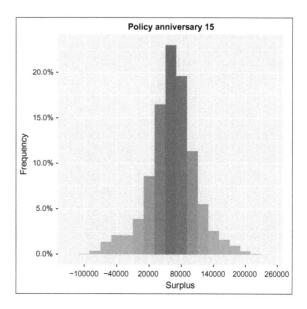

Fig. 7.15 Comparing
impacts: random fluctuations
vs random fluctuations and
systematic deviations

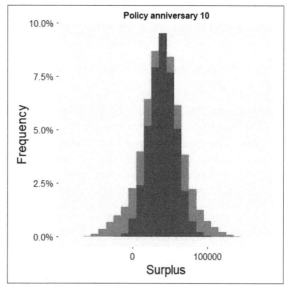

At time $t = 0$, the present value of profit or loss originated by an annuity policy is, of course, a random quantity, PL. Its expected value can be calculated as follows:

$$\overline{PL} = \Pi - b \, a_{x_0}^{[S]} \tag{7.11}$$

where S denotes a generic mortality scenario, and $a_{x_0}^{[S]}$ is calculated with a given discount rate. The single premium depends on the first-order technical basis adopted in the calculation, and in particular on the interest rate i (that is, a decision variable). Thus, we can more explicitly write:

$$\overline{PL}(i; S) = \Pi(i) - b \, a_{x_0}^{[S]} \tag{7.12}$$

We analyze the following problem: can a low interest rate i recover the worsening of the expected profit due to a mortality lower than expected?

Formally: $\Pi(i)$ and hence $\overline{PL}(i; S)$ (for any given S) increase as i decreases. Then, for a given S, find i' such that:

$$\overline{PL}(i; S) \geq 0 \quad \text{for } i \leq i' \tag{7.13}$$

(while the investment yield and the discount rate used to determine $a_{x_0}^{[S]}$ are assumed unchanged).

Example 7.10 The expected profit $N_0 \, \overline{PL}(i; S)$ originated by the annuity portfolio, as a function of the interest rate i, is plotted in Fig. 7.16, where three mortality scenarios are considered, i.e.,

Fig. 7.16 Expected profit as a function of the interest rate, in three different mortality scenarios

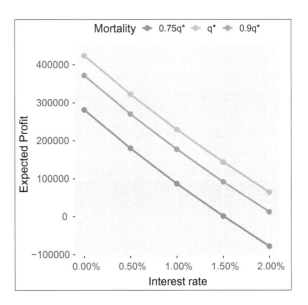

$$S_{(1)} = \{0.75\, q_x^*\}, \quad S_{(2)} = \{0.90\, q_x^*\}, \quad S_{(3)} = \{q_x^*\}$$

We recall that $S_{(3)}$ is assumed as the best-estimate scenario. We note, in particular, that the significant extra-survivorship implied by scenario $S_{(1)}$ results in a negative expected profit if $i = 0.02$, but can be faced if a lower interest rate is adopted in the premium calculation: indeed, $N_0\, \overline{PL}(i; S_{(1)}) > 0$ for $i < 0.015$. ∎

7.2 Solvency and Capital Allocation

In this section we first provide a definition of "portfolio default". Then, we focus on "solvency" and related shareholders' capital requirements. A stochastic approach and a deterministic one to the calculation of capital requirements will be presented. We specifically address a life annuity portfolio. For general issues regarding solvency and capital allocation, the reader is referred to Sects. 4.5 and 4.6.

7.2.1 Portfolio Default

A portfolio *default* occurs at time t if the amount of assets (i.e., the portfolio fund) is lower than the amount of liabilities, that is:

$$F_t^{[P]} < V_t^{[P]} \tag{7.14}$$

while $F_s^{[P]} \geq V_s^{[P]}$ for $s = 1, 2, \ldots, t - 1$.

Given the amount of the initial reserve $V_0^{[P]}$, the probability of default depends on the time horizon T referred to, and the availability of other resources, in particular the shareholders' capital M_0 initially allocated to the portfolio (assuming that no other capital flows occur up to time T). Then, the probability of default, assessed at time 0, is given by:

$$\varphi_0(T, M_0) = 1 - \mathbb{P}\left[\bigwedge_{t=1}^{T}(F_t^{[P]} \geq V_t^{[P]})\right] = 1 - \mathbb{P}\left[\bigwedge_{t=1}^{T}(M_t \geq 0)\right] \tag{7.15}$$

where $M_t = F_t^{[P]} - V_t^{[P]}$ denotes the capital assets, that is, the NAV of the portfolio (see Sect. 4.5.1).

7.2.2 Capital Requirement According to a Stochastic Model

Solvency has been introduced in Sect. 4.5.2. Formally, we can define the solvency in terms of the default probability. Hence, according to definition (7.15), the solvency capital requirement at time 0 is, for a given time horizon T, the amount M_0 such that:

$$\varphi_0(T, M_0) = \varepsilon \tag{7.16}$$

where ε denotes the accepted default probability.

An alternative approach to capital requirement is only based on the sign of M_T, that is at the end of the time frame. We define:

$$\psi_0(T, M_0) = 1 - \mathbb{P}\Big[M_T \geq 0\Big] \tag{7.17}$$

Then, the required capital is the amount M_0 such that:

$$\psi_0(T, M_0) = \varepsilon' \tag{7.18}$$

where ε' denotes the accepted probability.

Of course, requirements (7.16) and (7.18), for $\varepsilon = \varepsilon'$, coincide if $T = 1$, and the capital requirement is given by the absolute value of the VaR_ε of the probability distribution of M_1.

Remark Solving Eq. (7.16) or (7.18) is a very complicate computational problem, especially when other input random variables, such as yields from the assets backing the portfolio reserve, expenses, etc., which impact on the stochastic process $\{M_t\}$, are taken into account. For this reason, alternative approaches can be adopted.

- A number of *short-cut formulae* have been proposed and applied in insurance practice, which directly provide the capital requirement as a function of some quantities expressing the portfolio risk profile. The European standard formula for the SCR calculation, mentioned in Sect. 4.5.9, provides an example.
- Capital requirements can be determined following a deterministic approach, in particular stress testing in order to determine the amount of assets needed to face the impact of extreme events (see Sect. 7.2.3).

For more details regarding capital allocation for solvency purposes, see Sect. 4.5.9.

Example 7.11 We refer to the life annuity portfolio defined in Example 7.1. First, random fluctuations in annuitants' mortality around the best-estimate mortality assumption (see the parameters in Table 7.3 and the related markers in Table 7.4) are simulated. The default probabilities defined by Eq. (7.15), as functions of the time horizon T, are plotted in Fig. 7.17 for various values of the shareholders' capital M_0, expressed in terms of the ratio $M_0/V_0^{[P]}$. A modest capital allocation is required, in particular if short time horizons are considered, in order to keep the default probability at reasonable level. Higher capital allocations are required if systematic deviations

Fig. 7.17 Random
fluctuations: default
probability $\varphi_0(T, M_0)$

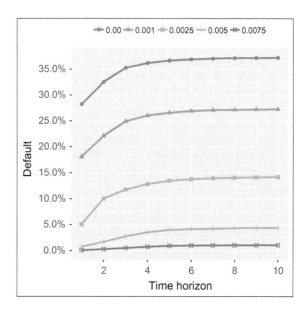

Fig. 7.18 Random fluct. +
systematic dev.: default
probability $\varphi_0(T, M_0)$

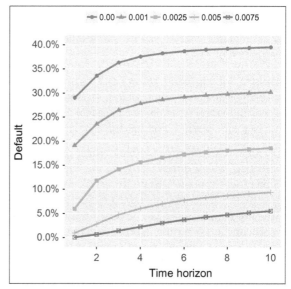

are also accounted for; see Fig. 7.18. However, default probabilities keep high, hence
suggesting an increase in capital allocation; see Fig. 7.19. Nonetheless, the impact
of the aggregate longevity risk becomes even more significant when a longer time
horizon is considered; see, for example, Fig. 7.12. To capture the impact of the aggregate longevity risk, a scenario-based approach can be more appropriate (see the next
section). ■

Fig. 7.19 Random fluct. +
systematic dev.: default
probability $\varphi_0(T, M_0)$

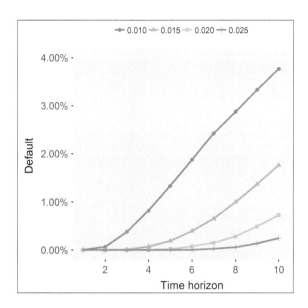

7.2.3 Capital Requirement Based on a Stress Scenario

A stress testing approach to capital requirement can be defined as follows.

1. Assume an immediate and permanent decrease in the annual probabilities of death.
2. Let α denote the decrease percentage (with respect to the best-estimate mortality), and define:

$$q_x^{*(\alpha)} = (1 - \alpha)\, q_x^*, \quad \text{for } x = x_0, x_0 + 1, \ldots \tag{7.19}$$

3. Let $M_0^{(\alpha)}$ denote the amount of shareholders' capital such that:

$$V_0^{[P]} + M_0^{(\alpha)} = \text{actuarial value of the benefits}$$
$$\text{according to probabilities } q_x^{*(\alpha)} \tag{7.20}$$

We note what follows.

- The stress scenario is defined in terms of "low" mortality and hence high probability of paying the life annuities for durations longer than anticipated in the prudential basis q_x.
- $M_0^{(\alpha)}$ is the required capital, thanks to which the insurer is able to meet the increased liability.
- The required capital is calculated by adopting a deterministic approach, as no assumption on the probability of a decrease in mortality is considered.

Fig. 7.20 Capital requirement $M_0^{(\alpha)}/V_0^{[P]}$ for different stress scenarios expressed by decrease percentage α

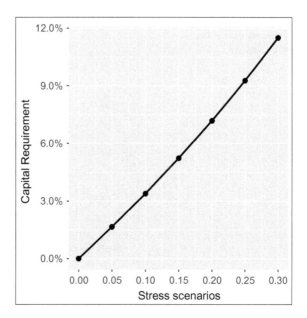

Example 7.12 We consider the portfolio referred to in the previous examples. Stress scenarios are defined in terms of an immediate and permanent decrease in the probability of death according to (7.19). The following values of α have been considered:

$$0.05, \ 0.10, \ 0.15, \ 0.20, \ 0.25, \ 0.30.$$

The corresponding capital requirements, in terms of the ratio $M_0^{(\alpha)}/V_0^{[P]}$ are plotted in Fig. 7.20. ∎

7.3 References and Suggestions for Further Reading

References on life annuities have been provided in Sect. 5.10. We only recall the book by Pitacco et al. (2009) which focuses on assessing and managing longevity risk in life annuity portfolios, and provides an extensive list of references.

The reader can find other bibliographic suggestions concerning the development of a life annuity product in Sect. 9.4.

Chapter 8
Sensitivity Testing for Long-Term Care Insurance Products

8.1 Outlook

Long-term care insurance (LTCI) covers (see Sect. 5.6) are rather recent products, in the framework of insurances of the person. It follows that specific biometric data are scanty. Pricing and reserving problems then arise because of difficulties in the choice of appropriate technical bases.

Different benefit structures imply different sensitivity degrees with respect to changes in biometric assumptions. Hence, an accurate sensitivity analysis can help in designing LTCI products and, in particular, in comparing stand-alone products to combined products, i.e., packages including LTCI benefits and other lifetime-related benefits.

Numerical examples (see Sects. 8.4.2 and 8.4.3) show, in particular, that the stand-alone cover is much riskier than all of the LTCI combined products that we have considered. As a consequence, the LTCI stand-alone cover is a highly "absorbing" product in terms of capital requirements.

8.2 The Biometric Model

In this section we first describe a simple multistate model which can be adopted to represent "states" and "transitions" related to a LTCI cover. We then introduce the biometric functions that are needed to assign a stochastic structure to the multistate model chosen.

This chapter is mainly based on the material presented in Pitacco (2016b), including numerical examples elaborated by Alice Petronio in her master thesis at the University of Trieste. The reader is referred to that paper for a more detailed presentation of sensitivity testing in long-term care insurance.

© Springer Nature Switzerland AG 2020

E. Pitacco, *ERM and QRM in Life Insurance*, Springer Actuarial,
https://doi.org/10.1007/978-3-030-49852-8_8

Fig. 8.1 The three-state
model

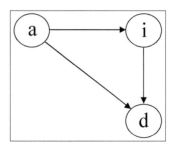

8.2.1 A Three-State Model

The multistate model we have adopted for the sensitivity testing simply consists of
three states and three possible transitions (see Fig. 8.1). The states are:

- a = active (i.e. healthy);
- i = incapacitated, or invalid (i.e. in LTC state);
- d = died;

and the transitions are as follows:

- a → i = disablement (i.e. entering the LTC state);
- a → d = death from the active state;
- i → d = death from the LTC state.

Hence, just one LTC state has been considered. Possibility of recovery has been
disregarded, because of the prevailing chronic character of LTC disability.

More complex multistate models can be considered, for example including two
or more disability states (corresponding to more or less severe health conditions) to
represent a degree-related benefit structure.

8.2.2 Biometric Functions

We refer to three-state model shown in Fig. 8.1. A Markov setting has been chosen,
thus probabilities only depend on the attained age and the current state, while no
inception-dependence is allowed for.

We define, for a healthy individual age x, the following one-year probabilities
(the traditional actuarial notation has been adopted):

p_x^{aa} = probability of being healthy at age $x + 1$;
p_x^{ai} = probability of being invalid at age $x + 1$;
q_x^{aa} = probability of dying before age $x + 1$ from state a;
q_x^{ai} = probability of dying before age $x + 1$ from state i;
q_x^{a} = probability of dying before age $x + 1$;
w_x = probability of becoming invalid (disablement) before age $x + 1$.

For an invalid individual (i.e. an individual in state LTC) age x, we consider the following one-year probabilities:

p_x^i = probability of being alive (and invalid) at age $x + 1$;
q_x^i = probability of dying before age $x + 1$.

The following relations obviously hold:

$$q_x^a = q_x^{aa} + q_x^{ai} \tag{8.1}$$

$$p_x^{aa} = 1 - q_x^{aa} - w_x \tag{8.2}$$

$$p_x^i = 1 - q_x^i \tag{8.3}$$

$$w_x = p_x^{ai} + q_x^{ai} \tag{8.4}$$

The (usual) approximation

$$q_x^{ai} \approx w_x \frac{q_x^i}{2} \tag{8.5}$$

has been assumed; in its turn, (8.5) implies:

$$p_x^{ai} \approx w_x \left(1 - \frac{q_x^i}{2} \right) \tag{8.6}$$

From the one-year probabilities, the following multi-year probabilities can be immediately derived:

$$_k p_x^{aa} = \prod_{h=0}^{k-1} p_{x+h}^{aa} \tag{8.7}$$

$$_k p_x^i = \prod_{h=0}^{k-1} p_{x+h}^i \tag{8.8}$$

$$_k p_x^{ai} = \sum_{h=1}^{k} \left[_{k-h} p_x^{aa} \; p_{x+k-h}^{ai} \; _{h-1} p_{x+k-h+1}^i \right] \tag{8.9}$$

Of course, we have: $_0 p_x^{aa} = 1$, $_0 p_x^i = 1$, and $_0 p_x^{ai} = 0$.

Example 8.1 We assume that:

q_x^{aa} is given by the third term of the first Heligman–Pollard law (see Eq. (4.9)), with parameter values as specified in Table 7.3;
w_x is expressed by a specific parametric law (see below);
$q_x^i = q_x^{aa} + $ extra-mortality (that is, an additive extra-mortality model is adopted).

The following assumption has been adopted for the one-year probability of disablement[1]:

---

[1] The assumption has been proposed by Rickayzen and Walsh (2002).

Table 8.1 Parameters of the function w_x

Parameter	Females	Males
a	0.0017	0.0017
b	1.0934	1.1063
c	103.6000	93.5111
d	0.9567	0.6591
e	n.a.	70.3002

$$w_x = \begin{cases} a + \dfrac{d-a}{1+b^{c-x}} & \text{for females} \\[2ex] \left(a + \dfrac{d-a}{1+b^{c-x}}\right)\left(1 - \dfrac{1}{3}\exp\left(-\left(\dfrac{x-e}{4}\right)^2\right)\right) & \text{for males} \end{cases} \qquad (8.10)$$

The relevant parameters are given in Table 8.1.

An additive extra-mortality model has been assumed to represent the mortality of disabled people. We have adopted the following formula[2]:

$$q_x^{i^{(k)}} = q_x^{[\text{standard}]} + \Delta(x, \alpha, k) \qquad (8.11)$$

with:

$$\Delta(x, \alpha, k) = \frac{\alpha}{1 + 1.1^{50-x}} \frac{\max\{k - 5, 0\}}{5} \qquad (8.12)$$

As regards the parameters and the related values, we note what follows.

- Parameter k expresses the LTC severity category, according to the OPCS[3] scale:
  $0 \le k \le 5$ denotes less severe LTC states, with no impact on mortality;
  $6 \le k \le 10$ denotes more severe LTC states, implying an extra-mortality.
  In the following calculations, we have assumed $k = 8$, that is, a severe LTC state (so that possibility of recovery can be disregarded). Hence, $q_x^i = q_x^{i^{(8)}}$ for all x.
- We have set $\alpha = 0.10$, as we have assumed $q^{[\text{standard}]} = q_x^{\text{aa}}$ (that is, the mortality of insured healthy people).[4]

From the previous assumptions, it follows:

$$\Delta(x, 0.10, 8) = \frac{0.06}{1 + 1.1^{50-x}}. \qquad (8.13)$$

■

---

[2] The formula has been proposed by Rickayzen and Walsh (2002).

[3] The acronym OPCS denotes the UK Office for Population Censuses and Surveys, former name of the Office for National Statistics (ONS).

[4] See Rickayzen (2007).

8.3 Actuarial Values. Premiums

The sensitivity testing presented in Sect. 8.4 is based on the analysis of single premiums as functions of the biometric assumptions. As is well known, according to the equivalence principle, the single premiums are given by the actuarial values of the benefits.

8.3.1 Actuarial Values

Let v denote the annual discount factor, $v = (1 + i)^{-1}$, where i denotes the interest rate adopted for discounting. We define the following actuarial values (i.e. expected present values). The usual actuarial notation has been adopted.

- Actuarial value, for a healthy individual age x (i.e. in state a), of a life annuity providing a benefit of 1 monetary unit per annum, payable at the policy anniversaries, while the individual is disabled (i.e. in state i):

$$a_x^{\text{ai}} = \sum_{j=1}^{+\infty} {}_{j-1}p_x^{\text{aa}}\, p_{x+j-1}^{\text{ai}}\, v^j\, \ddot{a}_{x+j}^{\text{i}} \tag{8.14}$$

- Actuarial value, for a disabled individual age $x + j$, of a life annuity providing a benefit of 1 monetary unit per annum, payable at the policy anniversaries, while the individual is disabled:

$$\ddot{a}_{x+j}^{\text{i}} = \sum_{h=j}^{+\infty} v^{h-j}\, {}_{h-j}p_{x+j}^{\text{i}} \tag{8.15}$$

- Actuarial value, for a disabled individual age $x + j$, of a temporary life annuity providing a benefit of 1 monetary unit per annum, payable at the policy anniversaries, while the individual is disabled:

$$\ddot{a}_{x+j:\overline{s}|}^{\text{i}} = \sum_{h=j}^{j+s-1} v^{h-j}\, {}_{h-j}p_{x+j}^{\text{i}} \tag{8.16}$$

- Actuarial value, for a healthy individual age x, of an n-year deferred life annuity of 1 monetary unit per annum, payable at the policy anniversaries, while the individual is healthy:

$$_{n|}\ddot{a}_x^{\text{aa}} = \sum_{j=n}^{+\infty} v^j\, {}_jp_x^{\text{aa}}. \tag{8.17}$$

8.3.2 Single Premiums

In the following sections, single premiums of four LTCI products are given (see Sect. 5.6.2 for the definitions of the relevant benefits):

1. Product P1: stand-alone LTC cover;
2. Product P2: LTC acceleration benefit in a whole-life assurance;
3. Product P3: LTC insurance package, including a deferred life annuity and a death benefit; in particular:

   a. Product P3a: Package a (fixed death benefit);
   b. Product P3b: Package b (decreasing death benefit);

4. Product P4: enhanced pension.

In the premium formulae, x_0 denotes the age at policy issue. In all the numerical calculations, we have assumed the interest rate $i = 0.02$, and hence $v = 1.02^{-1}$. The biometric functions specified in Example 8.1 have been assumed, referring to males. The premiums have been calculated by using the formulae given in Sect. 8.3.1.

Some comments follow, concerning the choice of the products to be analyzed.

- We first note that two "extreme" products have been included in the analysis, i.e. the stand-alone cover (product P1) and the whole-life assurance with LTCI as an acceleration benefit (product P2). While the former is a protection product which only aims at contributing to cover the LTC needs, the latter has an important savings component and a rather limited LTC coverage.
- The two remaining products either include a life annuity component (product P3) or are "derived" from a life annuity or pension product (product P4), thus aiming at covering the individual longevity risk.
- Although other products are sold on various insurance markets, the products we are addressing represent important LTCI market shares, and, at the same time, their simple structures ease the sensitivity analysis and the interpretation of the results.

8.3.3 Product P1: LTCI as a Stand-Alone Cover

According to the notation adopted in Sect. 8.3.1, the single premium, for an annual benefit b payable in arrears, is given by:

$$\Pi_{x_0}^{[P1]} = b\, a_{x_0}^{\text{ai}} \tag{8.18}$$

Example 8.2 Single premiums for various ages at policy issue are displayed in Table 8.2. ∎

Table 8.2 Product P1 (Stand-alone); Single premium $\Pi_{x_0}^{[P1]}$; $b = 100$

Age x_0	Single premium
40	480.431
50	513.544
60	516.465
70	473.732

8.3.4   Product P2: LTCI as an Acceleration Benefit

Refer to a whole life assurance with sum assured C, payable at the end of the year of death. The acceleration benefit consists in an s-year temporary LTC annuity with annual benefit C/s (this benefit could also be payable on a monthly basis, as mentioned in Sect. 5.6.2). The single premium, $\Pi_{x_0}^{[P2(s)]}$, of a whole life assurance with LTC acceleration benefit is given by:

$$\Pi_{x_0}^{[P2(s)]} = C \sum_{j=1}^{+\infty} {}_{j-1}p_{x_0}^{aa}\, q_{x_0+j-1}^{a}\, v^j$$

$$+ C \sum_{j=1}^{+\infty} {}_{j-1}p_{x_0}^{aa}\, p_{x_0+j-1}^{ai}\, v^j \left[\frac{1}{s} \ddot{a}_{x_0+j:s\rceil}^{i} + \sum_{h=1}^{s-1} \left(1 - \frac{h}{s}\right) {}_{h-1}p_{x_0+j}^{i}\, q_{x_0+j+h-1}^{i}\, v^h \right]$$

$$(8.19)$$

Conversely, the single premium for a (standard) whole life assurance is given by:

$$\Pi_{x_0}^{[WLA]} = C \sum_{j=1}^{+\infty} \left({}_{j-1}p_{x_0}^{aa}\, q_{x_0+j-1}^{a} + {}_{j-1}p_{x_0}^{ai}\, q_{x_0+j-1}^{i} \right) v^j \qquad (8.20)$$

Example 8.3 Single premiums for a (traditional) whole life assurance and for a whole life assurance with LTC acceleration benefit are shown in Table 8.3.   ∎

Table 8.3 Product P2 (Whole life assurance with LTC acceleration benefit); Single premiums $\Pi_{x_0}^{[WLA]}$ and $\Pi_{x_0}^{[P2(s)]}$; $C = 1\,000$

Age x_0	Whole life	Whole life with acceleration benefit				
	No accel.	$s = 1$	$s = 2$	$s = 3$	$s = 4$	$s = 5$
40	471.519	565.724	561.196	556.912	552.861	549.033
50	560.215	660.914	655.701	650.787	646.158	641.800
60	654.607	755.880	750.063	744.610	739.503	734.722

Table 8.4 Product P3a (Package a); Single premium $\Pi_{x_0}^{[P3a(x_0+n)]}$; $C = 1\,000$, $b' = 50$, $b'' = 150$

Age x_0	$x_0 + n = 75$	$x_0 + n = 80$	$x_0 + n = 85$
40	1 007.413	970.5772	955.9357
50	1 146.305	1 098.1236	1 078.9723
60	1 275.446	1 206.1263	1 178.5728
70	1 409.858	1 285.7893	1 236.4738

8.3.5   Product P3: LTCI in Life Insurance Package

For brevity, we only refer to Product P3a. The package provides the following benefits (see also Sect. 5.6.2):

- a life annuity with annual benefit b', deferred n years, while the individual is healthy (that is, an old-age life annuity, shortly ALDA);
- an LTC annuity with annual benefit b'';
- a death benefit C.

The single premium, $\Pi_{x_0}^{[P3a(x_0+n)]}$, is then given by:

$$\Pi_{x_0}^{[P3a(x_0+n)]} = b'\,_{n|}\ddot{a}_{x_0}^{aa} + b''\,a_{x_0}^{ai} + C\sum_{j=1}^{+\infty}\left(_{j-1}p_{x_0}^{aa}\,q_{x_0+j-1}^{a} + _{j-1}p_{x_0}^{ai}\,q_{x_0+j-1}^{i}\right)v^j$$

(8.21)

Example 8.4   Table 8.4 shows the single premiums for the product P3a.   ∎

8.3.6   Product P4: The Enhanced Pension

The single premium for a standard pension with annual benefit b is given by:

$$\Pi_{x_0}^{[SP(b)]} = b\sum_{j=1}^{+\infty}{_j}p_{x_0}^{a}\,v^j = b\sum_{j=1}^{+\infty}({_j}p_{x_0}^{aa} + {_j}p_{x_0}^{ai})\,v^j = b\,(a_{x_0}^{aa} + a_{x_0}^{ai})$$

(8.22)

The single premium for a pension with annual benefits b', b'', respectively paid if the annuitant is either healthy or in the LTC state, is given by:

$$\Pi_{x_0}^{[P4(b',b'')]} = b'\,a_{x_0}^{aa} + b''\,a_{x_0}^{ai}$$

(8.23)

In the case of an enhanced pension, we must have:

$$\Pi_{x_0}^{[P4(b',b'')]} = \Pi_{x_0}^{[SP(b)]}$$

(8.24)

Table 8.5 Product P4 (Enhanced pension); reduced benefit b', in order to obtain a given LTC benefit b'' ($b = 100$)

Age x_0	$\dfrac{\Pi_{x_0}^{[P4(b',b'')]}}{\Pi_{x_0}^{[SP(b)]}} =$	$b'' = 150$	$b'' = 200$	$b'' = 250$
60	1 761.478	79.259	58.517	37.776
65	1 522.646	75.824	51.649	27.473
70	1 278.444	70.565	41.130	11.695

and then:

$$b' a_{x_0}^{aa} + b'' a_{x_0}^{ai} = b (a_{x_0}^{aa} + a_{x_0}^{ai}) \tag{8.25}$$

From Eq. (8.25), given b, b'' we can calculate the reduced pension b'. Conversely, given b, b' we can find the uplifted pension b''.

Example 8.5 The reduced benefit b' which allows to obtain a given uplifted benefit b'' is displayed in Table 8.5, where a basic benefit $b = 100$ has been assumed. ∎

8.4 Sensitivity Testing

In this section we refer to the four LTCI products addressed in Sect. 8.3.2, and assess the sensitivity of single premiums with respect to assumptions concerning the probability of disablement and the extra-mortality.

8.4.1 A Formal Setting

Let $\Pi_{x_0}^{[PX]}(\delta, \lambda)$ denote the single premium of the LTCI product PX, with X = 1, 2, 3, according to the following assumptions:

- probability of entering the LTC state (i.e. probability of disablement) $\bar{w}_x(\delta)$, defined as follows:

$$\bar{w}_x(\delta) = \delta\, w_x \tag{8.26}$$

where w_x is given by Eq. (8.10), with parameters as specified in Table 8.1 for males;
- extra-mortality of people in the LTC state, defined as follows:

$$\bar{\Delta}(x; \lambda) = \lambda\, \Delta(x, 0.10, 8) = \frac{\lambda\, 0.06}{1 + 1.1^{50-x}} \tag{8.27}$$

(see Eq. (8.13)); it follows that the mortality of disabled people is given by:

$$q_x^i(\lambda) = q_x^{\text{aa}} + \bar{\Delta}(x; \lambda) \tag{8.28}$$

(of course, $\lambda = 0$ means absence of extra-mortality).

As regards the product P4, let $b'(\delta, \lambda)$ denote the amount of the reduced pension which meets a given uplifted pension b'' (for a given value of b), according to the assumptions expressed by δ and λ.

To ease the comparisons, we define for the LTCI products P1, P2 and P3 the "normalized" single premium, that is the ratio:

$$\rho_{x_0}^{[\text{PX}]}(\delta, \lambda) = \frac{\Pi_{x_0}^{[\text{PX}]}(\delta, \lambda)}{\Pi_{x_0}^{[\text{PX}]}(1, 1)} \tag{8.29}$$

whereas for the product P4, with given b and b'', we define the ratio:

$$\rho_{x_0}^{[\text{P4}]}(\delta, \lambda) = \frac{b'(1, 1)}{b'(\delta, \lambda)}. \tag{8.30}$$

8.4.2 Sensitivity Testing: Marginal Analysis

For all the products, we first perform a "marginal" analysis, by analyzing the behavior of the functions:

$$\Pi_{x_0}^{[\text{PX}]}(\delta, 1) \ \text{ for } X = 1, 2, 3; \quad b'(\delta, 1); \quad \rho_{x_0}^{[\text{PX}]}(\delta, 1) \ \text{ for } X = 1, 2, 3, 4$$

to assess the sensitivity with respect to the disablement assumption, and the functions:

$$\Pi_{x_0}^{[\text{PX}]}(1, \lambda) \ \text{ for } X = 1, 2, 3; \quad b'(1, \lambda); \quad \rho_{x_0}^{[\text{PX}]}(1, \lambda) \ \text{ for } X = 1, 2, 3, 4$$

to assess the sensitivity with respect to the extra-mortality assumption.

Some results of a "joint" sensitivity analysis are presented in Sect. 8.4.3.

Example 8.6 Results of sensitivity testing with respect to the disablement assumption are shown in Fig. 8.2. We immediately recognize the stand-alone LTCI product, i.e. product P1, as the one with the highest sensitivity with respect to the disablement assumption. The result is (rather) intuitive. We note, in particular, the dramatic impact of a possible underestimation of the probability of disablement, expressed by $\delta < 1$ (even excluding non-realistic underestimations, which could be represented by, say, $0 \leq \delta < 0.2$). ∎

Example 8.7 Results of sensitivity testing with respect to extra-mortality assumption are shown in Fig. 8.3. The stand-alone LTCI product, i.e. product P1, is the one with the highest sensitivity also with respect to the extra-mortality assumption. We

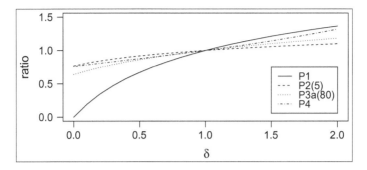

Fig. 8.2 Disablement assumption–Sensitivity testing

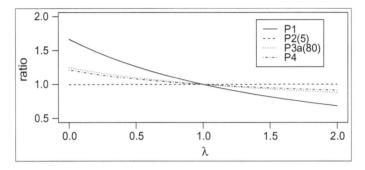

Fig. 8.3 Extra-mortality assumption—Sensitivity testing

note that, as P1 provides a living benefit, a safety loading could be included in the premium by underestimating the extra-mortality of disabled people. Conversely, the extra-mortality assumption has a very low impact in product P2. ∎

The following comments may help in interpreting the results plotted in the examples. First, we note that, intuitively, the actuarial values of payments linked to LTC increase as the probability of entering into LTC increases, and reduce as the extra-mortality from the LTC state increases. If an LTC benefit is added as a rider to accelerate insurance payments in a whole-life assurance, the certainty of the payment implies that the actuarial value is less sensitive relative to a similar LTC payment that is uncertain.

As regards the numerical values of the parameter δ, we in particular note that values lower than 1 (say, $\delta = 0.5$) can express a realistic estimate of the probability of becoming disabled, since only severe LTC states are considered to qualify the insured as eligible for LTCI benefits.

The parameter λ affects the extra-mortality of the disabled individuals. While parameter values close to 2 can represent a realistic estimate of the mortality of people in very severe health conditions, values lower than 1 lead to prudential actuarial valuations ($\lambda = 0$ representing, of course, the unrealistic absence of extra-mortality).

8.4.3 Sensitivity Testing: Joint Analysis

A joint sensitivity testing can produce various results of practical interest. It can be performed looking, in particular, at the surface which represents the behavior of the function

$$z = \Pi_{x_0}^{[PX]}(\delta, \lambda) \tag{8.31}$$

Example 8.8 Figure 8.4 shows the behavior of the function $z = \Pi_{50}^{[P3a(80)]}(\delta, \lambda)$. ∎

Sensitivity testing can be used to find, for the generic product PX and a given age x_0, the set of pairs (δ, λ) such that:

$$\rho_{x_0}^{[PX]}(\delta, \lambda) = \rho_{x_0}^{[PX]}(1, 1) = 1 \tag{8.32}$$

Equation (8.32) implies for products P1, P2, P3:

$$\Pi_{x_0}^{[PX]}(\delta, \lambda) = \Pi_{x_0}^{[PX]}(1, 1) \tag{8.33}$$

We note that Eq. (8.33) represents an "iso-premium" line: actually, all the pairs (δ, λ) which fulfill this equation lead to the same single premium.
For product P4, Eq. (8.32) implies:

$$b'(\delta, \lambda) = b'(1, 1) \tag{8.34}$$

Hence, the graphical representation of Eqs. (8.32) to (8.34) provides an insight into possible offset between, for example, an overestimation of the extra-mortality and an overestimation of the probability of entering the LTC state.

Fig. 8.4 Product P3a(80): z as a function of δ (disablement) and λ (extra-mortality)

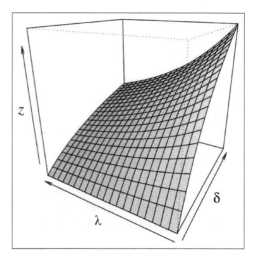

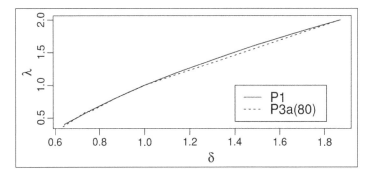

Fig. 8.5 Offset effect: iso-premium lines

Example 8.9 The iso-premium curves plotted in Fig. 8.5 show the offset possibility with reference to products P1 and P3a(80). We note that, thanks to the wide ranges chosen for both the parameters, the possibility appears effective also in the case of significant under- or over-estimation of the parameters. ∎

8.4.4 Profit Testing via Sensitivity Analysis

At the policy issue, the present value of the profit or loss originated by a LTCI product is, of course, a random quantity, PL. Its expected value, for the generic product PX, can be expressed as follows:

$$\overline{PL} = \Pi_{x_0}^{[\text{PX}]}(1, 1) - \Pi_{x_0}^{[\text{PX}]}(\delta, \lambda) \tag{8.35}$$

where (according to the notation defined in Sect. 8.4.1):

- $\Pi_{x_0}^{[\text{PX}]}(1, 1)$ denotes the single premium calculated with a given biometric basis;
- $\Pi_{x_0}^{[\text{PX}]}(\delta, \lambda)$ is the actuarial value of the LTC benefits, assessed by assuming that the biometric scenario is represented by the probabilities of disablement $\bar{w}_x(\delta)$ and the extra-mortality $\bar{\Delta}(x; \lambda)$ (see Eqs. (8.26) and (8.27), respectively).

Example 8.10 The impact of diverse disablement scenarios is shown in Table 8.6. A probability of entering the LTC state 50% lower than the one adopted in the premium calculation ($\delta = 0.5$) of course leads to a profit in all the three products, while a probability 50% higher ($\delta = 1.5$) implies a loss. Anyway, the lowest impacts can be found in the actuarial values of the whole life assurance with LTC acceleration benefit. ∎

Table 8.6 Actuarial values in diverse disablement scenarios

δ	$\Pi_{50}^{[P1]}(\delta, 1)$	$\Pi_{50}^{[P2(5)]}(\delta, 1)$	$\Pi_{50}^{[P3a(80)]}(\delta, 1)$
0.5	344.126	588.412	945.033
1.0	513.544	641.800	1 098.124
1.5	622.269	678.123	1 209.823

Table 8.7 Actuarial values in diverse extra-mortality scenarios

λ	$\Pi_{50}^{[P1]}(1, \lambda)$	$\Pi_{50}^{[P2(5)]}(1, \lambda)$	$\Pi_{50}^{[P3a(80)]}(1, \lambda)$
0.5	649.277	641.081	1 207.231
1.0	513.544	641.800	1 098.124
1.5	419.734	642.494	1 022.719

Example 8.11 The impact of diverse extra-mortality scenarios is shown in Table 8.7. We note, in particular, the very low impact on the actuarial values of the whole life assurance with LTC acceleration benefit. Conversely, an extra-mortality 50% lower than the one adopted in the premium calculation ($\lambda = 0.5$) implies a significant loss for the other products, while an extra-mortality 50% higher than the one adopted in the premium calculation ($\lambda = 1.5$) leads to a profit.   ∎

8.5 References and Suggestions for Further Reading

As already noted, the material in this chapter is mainly based on the achievements described in Pitacco (2016b), to which the reader should refer for a more detailed presentation of sensitivity testing for LTCI products.

Among the most recent publications on the actuarial aspects of LTCI, we suggest the book by Dupourqué et al. (2019). Actuarial models for pricing and reserving are presented by Haberman and Pitacco (1999), and Denuit et al. (2019).

Extra-mortality of disabled people, in the framework of heterogeneity modelling, is discussed by Pitacco (2019).

Biometric functions for implementing a multistate model for LTCI have been proposed by Rickayzen and Walsh (2002); OPCS data underpin the relevant parameter estimation. This topic has been progressed further by Rickayzen (2007). Results presented in these papers have been used in the present chapter.

Chapter 9
ERM and QRM in Product Development: An Example

9.1 Motivation and Outlook

To provide a more complete picture of ERM in the insurance field, encompassing all the phases of the RM process, we propose a simple case study: the development and the launch of special-rate life annuity products.

New Product Development (*NPD*) is a process by itself in all sectors of production, the insurance industry included. The NPD process brings a product idea to market. Many methodologies and techniques are available, which provide guidelines to implement the NPD process. Analyzing and comparing diverse methodologies is beyond the scope of this chapter. Then, we only focus on the Stage-Gate®[1] methodology, which is probably more easily applicable to life insurance. Although diverse sectors of production call for industry-specific processes, most of the features of the Stage-Gate methodology fit the needs of the NPD process in a number of sectors, the life insurance and life annuity business included.

Moreover, it is interesting to see how the flow of stages in the Stage-Gate process and the flow of phases in the RM process extended to the product design phase (see Sect. 5.2) can be combined. Any stage of the Stage-Gate process involves various departments and functions of the company, in particular relying on cross-functional teams. Hence, combining the two processes helps in stressing the multi-functional features of the RM phases, consistently with the ERM logic.

9.2 Product Development: Stage-Gate® and RM Process

In this section, we first provide a brief description of the Stage-Gate® system. We then focus on the development process for life insurance and life annuity products, in terms of both the Stage-Gate and the RM process.

---

[1] Stage-Gate® is a registered trademark of Stage-Gate Inc. (www.stage-gate.com).

© Springer Nature Switzerland AG 2020
E. Pitacco, *ERM and QRM in Life Insurance*, Springer Actuarial,
https://doi.org/10.1007/978-3-030-49852-8_9

9.2.1   The NPD Process According to the Stage-Gate® System

A Stage-Gate system is an effective way to implement the NPD process. According to the Stage-Gate original version,[2] the process starts with the *discovery* or "ideation phase", followed by five *stages* and five *gates*, and terminates with the *post-launch review*. In each stage, several parallel activities are performed by a cross-functional team. Each gate is a quality control checkpoint: a steering committee check the quality of the work done in the early stages and decides whether to go forward or to kill the project; action plan for the next stage is agreed.

The process consisting of the discovery, the five stages and the post-launch review can be described as follows.

- *Discovery.* In this phase, opportunities are analyzed and ideas are generated, in particular looking at the market, that is, customers' needs and products sold by competitors.
- *Stage 1.* This stage mainly consists of preliminary investigations, which aim at "scoping" the project. The company has to assess whether the product fits in its own market, to look at possible legislative constraints, to approximately assess the cost of the project, etc.
- *Stage 2.* A "business case" must be built, relying on research results, both market and technical. The product must be completely defined, and the project justified and planned. This stage should in particular include pricing of the product, evaluation of the likely market share, and profitability analysis.
- *Stage 3.* This stage can be denoted as "development stage". The product has to pass an alpha-test, that is, a test carried out in a lab environment, usually by internal employees of the organization.
- *Stage 4.* The product is tested on the market via release to a limited number of "users", thus has to pass a beta-test.
- *Stage 5.* Production, market launch and selling are included in this stage.
- *Post-launch review.* Post-launch reviews can be arranged according to a pre-defined timetable, e.g., three or four months after launch to assess projects strengths and weaknesses, and twelve months after launch to assess the product's performance. Various metrics can be used, in particular KPIs, for example A/E ratios to compare actual results to expected results.

All the Gates have the same structure, albeit different purposes depending on the place inside the sequence of stages. Basically, the structure of each gate can be summarized by the following points.

- The decision is based on the material delivered as the output of the previous stage.
- Decision criteria must be defined. Two basic criteria can be adopted:
  - "must-meet" criteria, leading to a go/kill decision;
  - "should-meet" criteria, leading to a score and hence to the assignment of a higher or lower priority to the project.

---

[2]See Cooper (2001).

- A decision must finally be taken, possibly suggesting improvements in the action plan.

While the original proposal focused on the 5-Stage structure, simplified structures have later been suggested,[3] which consist of a smaller number of stages via aggregation of some of the five stages. The choice of the number of stages, and hence of the relevant contents, of course depends on the complexity of the project.

Whatever the number of stages, the flow through the stages should not be considered as a strictly sequential path. In fact, one or more stages in the sequence can allow internal loops. For example, new information achieved while performing the stage activities may suggest appropriate adjustments and hence the repetition of some work inside the stage itself.

9.2.2 Combining Stage-Gate® and RM Process in Life Insurance Business

Combining the Stage-Gate methodology and the RM process structure can result in an effective approach to NPD in life insurance business.

We here refer to a 3-Stage implementation (instead of a more detailed 5-Stage implementation) of the Stage-Gate methodology. Three Gates are then required. The aggregation results in the following three Stages: Stage 1–2, Stage 3–4, Stage 5; see Fig. 9.1. Some comments follow, referring to the launch of a life insurance or life annuity product.

Starting on the Stage-Gate side, we note that the ideation of the product (phase *Discovery*) must comply with the targets of the company (*Objective setting* in the RM process: profit, market share, etc.). In particular, "Voice-of-Customer" research should play a substantial role by providing a significant input to the definition of the product features.

Stage 1–2 (*Scope* and *Business case*) should result in the complete definition of:

- the product "prototype" (*Product Design*), for which a tentative pricing must be proposed;
- the structure of the new line of business (or the extension of an existing line of business);
- the selection of appropriate distribution channels (e.g., agents, independent brokers, banks, internet).

A business plan must be constructed, by projecting the expected cash-flows for the next five years, say. Scenario variables and decision variables must be identified (see Sect. 4.4.3), and appropriate assessments must be performed via what-if-analysis, in particular scenario testing and sensitivity testing (see Sect. 4.4.6). Main purposes are as follows:

---

[3]See, for example, Cooper (2008, 2010).

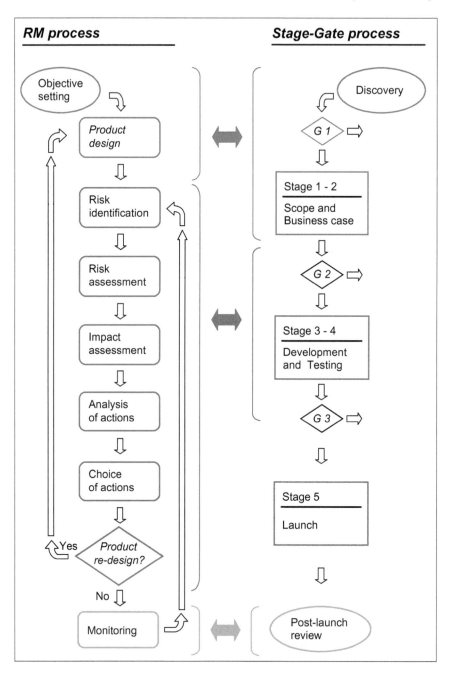

Fig. 9.1 New Product Development: Framing the RM process into Stage-Gate®

– assessing the time-profile of the profits;
– quantifying capital requirements due to new business strain;
– investigating what volumes are required to cover expenses.

Significant technical, financial and market issues characterize Stage 3–4 (*Development* and *Testing*). The core of the RM process (from *Risk identification* to *Choice of actions*) is involved. In particular, we note what follows.

- At this point, all the policy conditions must be defined (e.g. surrender if applicable, settlement options regarding the payment of death benefit if any, contract term extension, paid-up option, etc.; see relevant sections in Chap. 5).
- Pricing and reserving should rely on rigorous premium principles and sound statistical bases (e.g., mortality tables, disability tables); expense loadings should be quantified. Reinsurer's advice can provide significant help in the choice of technical bases.
- Underwriting procedures should be defined according to the type of insurance product, and the distribution channels chosen.
- Appropriate risk and impact assessments must be performed, and RM actions should then be chosen (reinsurance, capital allocation, etc.). The role of QRM clearly appears.
- A market test can be simulated by delivering the product to a selected set of agents, in order to obtain judgements on the marketability of the product.
- The result of the above actions may be either moving to the next Stage (*Launch*) and hence the delivery of the product to the market, or its redesign (*Product redesign*, in the RM process) and then the reiteration of the various phases.
- In the case of delivery of the product to the market (*Launch*), the NPD process according to the Stage-Gate methodology terminates with the final review (*Post-launch review*). Possibly more reviews can be arranged according to a predefined timetable. Conversely, in the RM process applied to a (either new or already existing) line of business the review phase (*Monitoring*) is a recurrent phase whose timing must be defined according to the objects of the review itself (see Sect. 3.6).

The possible redesign of the product witnesses the non necessarily sequential feature of the path through the Stages. We also note that, in the case of redesign, the new product design again involves Stage 1–2 of the process, as the construction of a new business case might be required if the redesign implies important changes in the product (e.g., changes in the policy conditions, introduction or removal of rider benefits).

9.3 The NPD Process for Special-Rate Life Annuities

In this section we present an application of the Stage-Gate® process combined with the RM process, addressing the development of special-rate annuities. We first recall the main features of this product. We then focus on those Stages and RM phases which are particularly involved in the NPD process because of the specific features of the product itself.

9.3.1 The Product and Its Motivation

As noted in Sect. 5.5, (standard) life annuities are attractive mainly for healthy people. Premium rates are consequently kept high. Hence, a large portion of potential annuitants are out of reach of insurers (see Fig. 9.2).

Lowering the premium rate would of course raise the attractiveness of the life annuity product (see Fig. 9.3), but would result in a more heterogeneous portfolio and in a likely underpricing in particular for healthy annuitants. Such a solution should then be considered unrealistic.

In order to expand their business, some insurers have recently started offering better annuity rates to people whose health conditions are worse than those of likely buyers of (standard) life annuities. *Special-rate life annuity* (or *underwritten life annuities*) products have then been designed.

The following special-rate annuities are sold in several markets.

1. The underwriting of a *lifestyle annuity* takes into account smoking and drinking habits, marital status, occupation, height and weight, blood pressure and cholesterol levels. These factors might result, to some extent, in a shorter life expectancy.
2. The *enhanced life annuity* pays out an income to a person with a reduced life expectancy, in particular because of a personal history of medical conditions. Of course, the "enhancement" in the annuity benefit (compared to a standard-rate life annuity, same premium) comes, in particular for this type of annuity, from the use of a higher mortality assumption.

Fig. 9.2 Potential annuitants population and standard annuity portfolio

Fig. 9.3 Potential annuitants population and (unrealistic) better-rate annuity portfolio

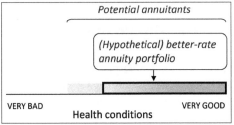

Fig. 9.4 Potential annuitants population and annuity portfolio also consisting of three special-rate annuity sub-portfolios

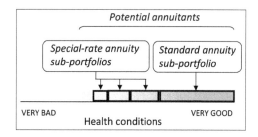

Fig. 9.5 Curves of deaths for different life annuity sub-portfolios

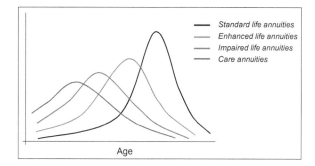

3. The *impaired-life annuity* pays out a higher income than an enhanced life annuity, as a result of medical conditions which significantly shorten the life expectancy of the annuitant (e.g., diabetes, chronic asthma, cancer, etc.).
4. Finally, *care annuities* are aimed at individuals with very serious impairments or individuals who are already in a senescent-disability (or long-term care) state.

Thus, moving from type 1 to type 4 results in progressively higher mortality assumptions, shorter life expectancy, and hence, for a given single premium amount, in higher annuity benefits. In particular, regarding annuities of types 3 and 4, the underwriting process results in classifying the applicant as a *substandard risk*, because of ascertainment of significant extra-mortality. For this reason, annuities of types 3 and 4 are sometimes named *substandard annuities*.

Of course, an insurer can decide to offer a more limited set of alternatives, say two or three different premium rates (see Fig. 9.4).

Premium rates must be determined according to the probability distribution of the lifetime of individuals belonging to the various sub-portfolios. Four curves of deaths (or probability density functions in time-continuous setting) are sketched in Fig. 9.5. The worse the health conditions, the smaller the modal age at death (as well as the expected lifetime), but the higher the variance of the lifetime distribution. The latter aspect is due to the mix of possible pathologies leading to each specific individual classification. A higher degree of (partially unobservable) heterogeneity follows, inside each sub-portfolio of special-rate annuities.

9.3.2   The NPD Process: Specific Issues

To expand its life annuity portfolio, by reaching a larger number of potential annuitants, a company aims at launching classes of special-rate, single-premium life annuities (*Discovery* stage). The financial structure is a "conventional" one (that is, non-unit-linked), possibly with profit participation.

Although similar products are already sold in various markets, the features of special-rate life annuities must be carefully analyzed, and a number of business characteristics must be considered. Then, a NPD process is needed, involving many functions and processes of the company, and hence a cross-functional team.

In the following, critical issues are singled-out, and reference to the relevant Stages and RM process phases is stressed. The 3-Stage structure is adopted.

The company in particular aims at the following targets (*Objective setting*):

- increasing its market share;
- profitability.

Stage 1–2 (*Scope* and *Business Case*), as noted in Sect. 9.2.2, must lead to a provisional definition of the product (*Product design*). Developing and selling special-rate annuity products can be considered an extension of the existing line of business in which standard life annuity are placed. Nevertheless, the features of special-rate life annuities suggest specific choices, in particular regarding distribution channels. While direct marketing (e.g., via internet) must be excluded because of the underwriting requirements, tied agents employed by the insurer can provide an appropriate solution. As already stressed (see Sect. 9.2.2), a business plan must be developed to assess the profitability of the new product and to quantify capital requirements.

Stage 3–4 (*Development* and *Testing*) involves important technical, financial and market issues. In the next sections we focus on two issues, which are typically connected to the specific feature of the product.

Given the target of special-rate life annuities, mortality higher than that of people purchasing standard life annuities must be ascertained via *underwriting requirements*. The underwriting process can be arranged in a number of ways, as explained in Sect. 9.3.3.

All life annuity providers are exposed to longevity risk, and in particular to its systematic component, that is, the aggregate longevity risk. In special-rate annuity business, a more complex risk profile is due to a significant uncertainty in choosing the lifetime distributions because of scarcity of data. Moreover, some degree of uncertainty also affects the results of the underwriting process. Relevant impacts on results of interest should then be assessed. QRM methodologies play a preponderant role. Some aspects will be addressed in Sect. 9.3.4.

As regards policy conditions, we note what follows.

- Surrender is not allowed, as in all life annuity products, to avoid adverse selection.
- A guarantee period (5 years, say), as well as a capital protection (or money-back) option which adds a death benefit to the life annuity product (see Sect. 5.5.1), may make the annuity product more attractive. However, it is worth noting that the

mortality assumptions underlying special-rate annuities make these rider benefits more expensive than similar riders in standard life annuities.

- Profit participation mechanisms can be implemented, possibly including also mortality profits.
- A special-rate life annuity cannot be sold as a last-survivor annuity (again, see Sect. 5.5.1) if, as usual, health conditions are only assessed for one beneficiary.

Stage 5 (*Launch*) and the final review (*Post-launch review*) conclude the NPD process. More final reviews can be scheduled (3, 6 and 12 months after launch, say) to assess the product performance on the market. Conversely, much longer periods are needed for an effective review of the statistical bases adopted (*Monitoring*), because trends slowly emerge over time, and specific data are anyway scanty.

We finally note that the launch of special-rate annuities can impact on the standard life annuity portfolio, In particular:

- future sales of standard annuities might decrease because of a shift of eligible customers towards special-rate annuities (the so called "cannibalization" effect);
- consequently, annuitants' mortality in the standard annuity portfolio might decrease, hence requiring a revision of the biometric assumptions underlying the standard premium rates.

These aspects should also be considered in the NPD process, notably in Stage 3–4 and 5.

9.3.3 Approaches to Underwriting

Underwriting for special-rate life annuities can be implemented in a number of ways, and several classification can be conceived.[4] It is interesting to focus on:

(1) what risk factors can be chosen as rating factors, besides annuitant's age and gender (if permitted by the local current legislation);
(2) how many rating factors are actually accounted for in the underwriting process of a given special-rate annuity;
(3) how many rating classes, that is, how many different annuity rates, are defined.

As regards (1), we note that higher mortality, and then lower life expectancy, may generally be due to the following causes.

(1a) The individual health, and in particular the presence of some past or current *disease*, clearly impacts on the mortality pattern.
(1b) The applicant's *lifestyle* (e.g. smoking and drinking habits, sedentary life, etc.) can cause higher mortality.

---

[4]What follows is mainly based on the classifications proposed by Rinke (2002). Differences in terminology can however be detected. In particular, the expression "enhanced annuity" is used, in that paper, as a synonym for underwritten or special-rate life annuity.

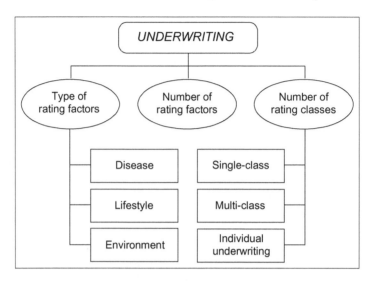

Fig. 9.6 Approaches to underwriting for special-rate life annuities

(1c) The *environment* in which the applicant lives might also impact on his/her mortality, and hence socio-geographical risk factors can be accounted for.

On the one hand, the higher the number of rating factors (see point (2) above), the more complex is the underwriting process; on the other, the higher the number of rating classes (see point (3)) the better is the fitting of the individual risk profile.

As regards the number of rating classes, the following classification reflects alternative pricing schemes that can be adopted in the insurance practice.

(3a) When a *single-class* underwriting scheme is adopted, one or just a few rating factors are accounted for, and the underwriting results in a yes/no answer. If yes, a given annuity rate, higher than the "standard" one, is applied. An example is given by the smoking habits. The portfolio then consists of a standard annuity sub-portfolio and a special-rate annuity sub-portfolio.

(3b) The *multi-class* underwriting scheme can be implemented either considering just one rating factor with several possible values, or more rating factors in which case each combination of values yields an annuity rate. More than one special-rate sub-portfolio is the result of this scheme.

(3c) The *individual underwriting* allows to use all the available information about the individual, so that the annuity rate can be tailored on the applicant's characteristics. Also this approach will result in diverse special-rate sub-portfolios.

The above classifications are sketched in Fig. 9.6.

9.3.4 The Biometric Risk Profile

Diverse biometric risk components combine in a special-rate annuity sub-portfolio (see definitions provided in Sect. 4.1.2). In detail:

1. mortality random fluctuations;
2. mortality systematic deviations, thus aggregate longevity risk, in particular due to:

   a. uncertainty in mortality level of beneficiaries of special-rate annuities;
   b. uncertainty in future longevity trend, as in all life annuity products.

Further, the diversity in pathologies affecting individuals classified in special-rate sub-portfolio implies risk due to:

3. heterogeneity in mortality inside each sub-portfolio.

The impact assessment of the risk profile of special-rate annuities call for appropriate calculation models. A sound QRM approach is then needed. Various RM actions are then required to manage the above risk components.

As regards the impact of mortality random fluctuations (component 1), the assessment can be performed by adopting a simple stochastic simulation model, as described in Sect. 7.1.6. Assessing the impact of systematic deviations (components 2.a and 2.b) can be approached, for example, defining a set of diverse scenarios, as suggested in Sect. 7.1.7. Further, after assigning a probability distribution on the scenario space, a double stochastic approach can be adopted to jointly assess the impact of the two risk components, as described in Sect. 7.1.8.

The choice of the "baseline" mortality assumptions for the various special-rate sub-portfolios is, of course, a critical issue in the NPD process. In this regard, reinsurers can provide significant advice.

As regard RM actions, we note what follows.

- Capital allocation is of course required. Given the particular features of special-rate annuity sub-portfolios, internal assessments in the framework of ORSA can yield appropriate capital requirement.
- The coexistence of the various risk components (1, 2.a and 2.b) calls for reinsurance solutions alternatively arranged as follow (see Sects. 4.6.5 and 4.6.6):

  – traditional reinsurance transfer to the reinsurer, which will then resort to ARTs;
  – ART from insurer to reinsurer, for example via a swap arrangement.

Finally, the presence of risk due to heterogeneity in mortality can be mitigated by restricting the range of pathologies which entitle to a special-rate annuity, thus making the relevant sub-portfolio more homogeneous.

9.4   References and Suggestions for Further Reading

A number of papers and technical reports deal with product development, product design, and related technical issues. For example, a survey on product development practices adopted by life insurance and annuity companies in the US and Canada markets is provided by SOA (2017). A classical reference for understanding critical points in product innovation in life insurance is given by Rudelius and Wood (1970). An overview on product design in life insurance is provided, for example, by Falconer (2003). Starting from critical aspects of the demand for life insurance, Kirova and Steinmann (2013) single out feasible strategies to improve people understanding of the nature of risks and the role of life insurance.

The Stage-Gate® system has been adopted by many leading companies to implement the product innovation process; see Cooper (2001, 2008, 2010). In particular, Sect. 9.2.1 of this book is mainly based on Cooper (2001), while implementation in Sect. 9.2.2 relies on simplifications in the Stage-Gate process suggested by Cooper (2008, 2010). A discussion on product development process in life insurance and, in particular, on the application of the Stage-Gate® method is available in SOA (2003).

References to general issues regarding life annuities have been provided in Sect. 5.10. We only recall papers and reports specifically addressing special rate life annuities: Ainslie (2000), Drinkwater et al. (2006), Ridsdale (2012) and Rinke (2002). In particular, the impact of risk classification on the structure of life annuity portfolios is analyzed by Gatzert et al. (2012b), Hoermann and Russ (2008) and Olivieri and Pitacco (2016).

An interesting analysis of market issues related to special-rate annuities is presented by Gatzert and Klotzki (2016), where barriers on the demand side and the supply side are specifically addressed.

Observable and unobservable heterogeneity in mortality is addressed by Pitacco (2019), where an extensive list of references is also provided.

References

Ainslie, R. (2000). Annuity and insurance products for impaired lives. Working Paper. Presented to the Staple Inn Actuarial Society.

Beasley, M. S., Branson, B. C., & Hancock, B. V. (2010). Developing key risk indicators to strengthen enterprise risk management. Technical report, COSO - Committee of Sponsoring Organizations of the Treadway Commission. https://www.coso.org/Documents/COSO-KRI-Paper-Full-FINAL-for-Web-Posting-Dec110-000.pdf.

Bellis, C., Lyon, R., & Klugman, S., (Eds.). (2010). *Understanding actuarial management: The actuarial control cycle* (2nd ed.). The Institute of Actuaries of Australia and the Society of Actuaries.

Black, K., & Skipper, H. D. (2000). *Life and health insurance*. New Jersey: Prentice Hall.

Blake, D., Cairns, A., & Dowd, K. (2006). Living with mortality: Longevity bonds and other mortality-linked securities. *British Actuarial Journal, 12*(1), 153–197.

Blake, D., Cairns, A., Dowd, K., & Kessler, A. (2019). Still living with mortality: The longevity risk transfer market after one decade. *British Actuarial Journal, 24*(E1),

Bohnert, A., Gatzert, N., Hoyt, R. E., & Lechner, P. (2017). The relationship between enterprise risk management, value and firm characteristics based on the literature. *ZVersWiss, 106*, 311–324.

Bohnert, A., Gatzert, N., Hoyt, R. E., & Lechner, P. (2019). The drivers and value of enterprise risk management: Evidence from ERM ratings. *The European Journal of Finance, 25*(3), 234–255.

Booth, P., Chadburn, R., Haberman, S., James, D., Khorasanee, Z., Plumb, R. H., et al. (2005). *Modern actuarial theory and practice*. London: Chapman & Hall/CRC.

Boyd, S. R., Moolman, J. A., & Nwosu, N. J. (2016). Risk reporting and key risk indicators. A case study analysis. Technical report, North Carolina State University - Poole College of Management - Enterprise Risk Management Initiative. https://erm.ncsu.edu/az/erm/i/chan/library/ERM_KRI_Case_Study_FINAL.pdf.

Boyle, P., & Hardy, M. (2003). Guaranteed annuity options. ASTIN. *Bulletin, 33*(2), 125–152.

Branson, B. C. (2015). Reporting key risk information to the board of directors. Technical report, North Carolina State University - Poole College of Management - Enterprise Risk Management Initiative. https://erm.ncsu.edu/az/erm/i/chan/library/2015-erm-reporting-key-risk-information-to-board-directors.pdf.

Brighenti, L. (2015). Assicurazioni miste rivalutabili: Analisi di garanzie alternative di rendimento. Master's thesis, Università di Trieste.

Calandro, J. (2006). An enterprise approach to insurance risk management. White paper, IBM Corporation - IBM Global Services, U.S.A. https://www-304.ibm.com/easyaccess/fileserve?contentid=96554.

CAS. (2003). *Overview of enterprise risk management*. Casualty Actuarial Society - Enterprise Risk Management Committee. https://www.casact.org/area/erm/overview.pdf.

© Springer Nature Switzerland AG 2020
E. Pitacco, *ERM and QRM in Life Insurance*, Springer Actuarial,
https://doi.org/10.1007/978-3-030-49852-8

Chan, C. Y. (2014). The impact of gender-neutral pricing on the life insurance industry. Master's thesis, Cass Business School, City University London. http://publications.scor.com/actuarial_prize/2015_uk_ChanCho-Yeung.pdf.

Chapman, R. J. (2006). *Simple tools and techniques for enterprise risk management.* New Jersey: Wiley.

Chou, T., Fulbright, J., Irwin, C., & Patch, M. (2020). The art and science of key risk indicators: A case study analysis. Technical report, North Carolina State University - Poole College of Management - Enterprise Risk Management Initiative. https://erm.ncsu.edu/library/article/the-art-science-of-key-risk-indicators-a-case-study-analysis.

Coggins, M., Dexter, N., Kemp, M., & Oost, J. (2016). Own risk and solvency assessment (ORSA). In *IAA Risk Book, chapter 10.* International Actuarial Association. https://www.actuaires.org/LIBRARY/Papers/RiskBookChapters/Ch10_ORSA_8March2016.pdf.

Cooper, R. G. (2001). *Winning at new products: Accelerating the process from idea to launch* (3rd ed.). New York: Perseus Books.

Cooper, R. G. (2008). The stage-gate idea-to-launch process - Update, what's new and NexGen systems. *Journal of the Product Innovation Management, 25*(3), 213–232.

Cooper, R. G. (2010). The stage-gate idea to launch system. In J. N. Sheth & N. K. Malhotra (Eds.), *Wiley International Encyclopedia of Marketing.* Hoboken: Wiley Ltd.

COSO. (2004). *Enterprise risk management - integrated framework (Executive summary).* Committee of Sponsoring Organizations of the Treadway Commission. https://www.coso.org/Documents/COSO-ERM-Executive-Summary.pdf.

Crouhy, M., Galai, D., & Mark, R. (2001). *Risk management.* New York: McGraw-Hill.

Curry, C., & O'Connell, A. (2004). An analysis of unisex annuity rates. Equal Opportunities Commission, working paper series, No. 16. www.eoc.org.uk/research.

Denuit, M., Lucas, N., & Pitacco, E. (2019). Pricing and reserving in LTC insurance. In E. Dupourqué, F. Planchet, & N. Sator (Eds.), *Actuarial aspects of long term care, Springer Actuarial* (pp. 129–158). Berlin: Springer.

Dickinson, G. (2001). Enterprise risk management: Its origins and conceptual framework. *The Geneva Papers on Risk and Insurance, 26*(3), 360–366.

Doherty, N. (2000). *Integrated risk management: Techniques and strategies for managing corporate risk.* New York: McGraw Hill.

Dowd, K. (1998). *Beyond value at risk. The new science of risk management.* Hoboken: Wiley.

Drinkwater, M., Montminy, J. E., Sondergeld, E. T., Raham, C. G., & Runchey, C. R. (2006). Substandard annuities. Technical report, LIMRA International Inc. and the Society of Actuaries, in collaboration with Ernst & Young LLP. https://www.soa.org/Files/Research/007289-Substandard-annuities-full-rpt-REV-8-21.pdf.

Dupourqué, E., Planchet, F., & Sator, N. (Eds.). (2019). *Actuarial aspects of long term care. Springer actuarial.* Berlin: Springer.

Edwards, M. (2008). The last post. *The Actuary, 2008*(9), 30–31. http://www.theactuary.com/archive/2008/09/.

Eves, M., Fritsch, A., & Müller, E. (2015). Non-proportional reinsurance. In *IAA Risk Book, chapter 6.* International Actuarial Association. https://www.actuaries.org/LIBRARY/Papers/RiskBookChapters/Ch6_Non-proportional_Reinsurance_2015-08-28.pdf.

Falconer, C. (2003). Designing life insurance products - A broad overview. *AIO Life Seminar.*

Farrell, M., & Gallagher, R. (2015). The valuation implications of enterprise risk management maturity. *The Journal of Risk and Insurance, 82*(3), 625–657.

Gatzert, N., Holzmüller, I., & Schmeiser, H. (2012a). Creating customer value in participating life insurance. *Journal of Risk and Insurance, 79*(3), 645–670.

Gatzert, N., & Klotzki, U. (2016). Enhanced annuities: Drivers of and barriers to supply and demand. *The Geneva Papers on Risk and Insurance - Issues and Practice, 41*(1), 53–77.

Gatzert, N., Schmitt-Hoermann, G., & Schmeiser, H. (2012b). Optimal risk classification with an application to substandard annuities. *North American Actuarial Journal, 16*(4), 462–486.

Goford, J. (1985). The control cycle: Financial control of a life assurance company. *Journal of the Staple Inn Actuarial Society, 28*, 99–114.

Gorvett, R. (2006). The role of auditing in the ERM process. SOA annual meeting, Chicago, October 2006. www.math.uiuc.edu/~gorvett/present/soaerm.ppt.

Gribble, J. D. (2003). Actuarial practice and control: Objectives and capabilities. Research paper no. 105, Centre for Actuarial Studies, Department of Economics, The University of Melbourne, Victoria (Australia). https://minerva-access.unimelb.edu.au/handle/11343/33755.

Gutterman, S. (2016). Distribution risks. In *IAA Risk Book, chapter 9*. International Actuarial Association. https://www.actuaires.org/LIBRARY/Papers/RiskBookChapters/Ch9_Distribution_Risks_8March2016.pdf.

Gutterman, S. (2017). Risk and uncertainty. Quantification, communication and management. In *IAA Risk Book, chapter 17*. International Actuarial Association. https://www.actuaires.org/LIBRARY/Papers/RiskBookChapters/Ch17_Risk_and_Uncertainty_6June2017.pdf.

Haberman, S. (1996). Landmarks in the history of actuarial science (up to 1919). Actuarial research paper no. 84, Faculty of Actuarial Science & Insurance, City University, London. Available at: http://openaccess.city.ac.uk/2228/1/84-ARC.pdf.

Haberman, S., & Olivieri, A. (2014). *Risk classification/life. In Wiley StatsRef: Statistics reference online*. Hoboken: Wiley.

Haberman, S., & Pitacco, E. (1999). *Actuarial models for disability insurance*. Boca Raton: Chapman & Hall/CRC.

Hardy, M. R. (2003). *Investment guarantees: Modeling and risk management for equity-linked life insuranxe*. Hoboken: Wiley.

Harrington, S. E., & Niehaus, G. R. (1999). *Risk management and insurance*. New York: Irwin/McGraw-Hill.

Heligman, L., & Pollard, J. H. (1980). The age pattern of mortality. *Journal of the Institute of Actuaries, 107*(1), 49–80.

Heling, M., & Holder, S. (2013). The value of interest rate guarantees in participating life insurance contracts: Status quo and alternative product design. *Insurance: Mathematics and Economics, 53*(3), 491–503.

Hoermann, G., & Russ, J. (2008). Enhanced annuities and the impact of individual underwriting on an insurer's profit situation. *Insurance: Mathematics & Economics, 43*(1), 150–157.

Howes, T., Perrott, G., Selby, S., & Sherwood, D. (2019). Governance of models for insurance companies. In *IAA Risk Book, chapter 15*. International Actuarial Association. https://www.actuaires.org/IAA/Documents/CMTE_IRC/RiskBook_project/Revised_Chapters/Ch15_Governance_of_Models_21May2019_FinalRevised.pdf.

Hoyt, R. E., & Liebenberg, A. P. (2015). Evidence of the value of enterprise risk management. *Journal of Applied Corporate Finance, 27*(1), 41–48.

Hughes, M. (2012). Preferred risk in life insurance. SCOR inFORM, SCOR Global Life. https://www.scor.com/en/files/preferred-risk-life-insurance.

IAA. (2004). A global framework for insurer solvency assessment. International Actuarial Association - Insurer Solvency Assessment Working Party. http://www.actuaries.org/LIBRARY/Papers/Global_Framework_Insurer_Solvency_Assessment-members.pdf.

IAA. (2009). Note on enterprise risk management for capital and solvency purposes in the insurance industry. International Actuarial Association. http://www.actuaries.org/CTTEES_FINRISKS/Documents/Note_on_ERM.pdf.

IAA. (2010a). Comprehensive actuarial risk evaluation (CARE). International Actuarial Association - Enterprise and Financial Risk Committee. http://www.actuaries.org/CTTEES_FINRISKS/Documents/CARE_EN.pdf.

IAA. (2010b). Note on the use of internal models for risk and capital management purposes by insurers. International Actuarial Association - Solvency Subcommittee of the IAA Insurance Regulation Committee. http://www.actuaries.org/CTTEES_SOLV/Documents/Internal_Models_EN.pdf.

IAA. (2011). Note on enterprise risk management for pensions. International Actuarial Association - Pensions and Employee Benefits Committee. http://www.actuaries.org/LIBRARY/Papers/Note_ERM_Pensions_EN.pdf.

IAA. (2013). Stress testing and scenario analysis. International Actuarial Association - Insurance Regulation Committee. https://www.actuaries.org/CTTEES_SOLV/Documents/StressTestingPaper.pdf.

IAA. (2016). Actuarial aspects of ERM for insurance companies. International Actuarial Association - Enterprise and Financial Risk Committee. http://www.actuaries.org/CTEES_FINRISKS/Papers/ActuarialAspectsofERMforInsuranceCompanies_January2016.pdf.

IAA. (2017). Long-term care: An actuarial perspective on societal and personal challenges. International Actuarial Association - Population Issues Working Group. https://www.actuaires.org/LIBRARY/Papers/PIWG_LTC_Paper_April2017.pdf.

IAA. (n.d.). *IAA Risk Book*. International Actuarial Association. https://www.actuaries.org/iaa/IAA/Publications/iaa_riskbook/IAA/Publications/risk_book.aspx?hkey=1bb7bce0-2c43-41df-9956-98d68ca45ce4.

IAIS. (2019). Insurance core principles and common framework for the supervision of internationally active insurance groups. International Association of Insurance Supervisors. https://www.iaisweb.org/page/supervisory-material/insurance-core-principles.

IIA. (2009). IIA position paper: The role of internal auditing in enterprise-wide risk management. The Institute of Internal Auditors. https://na.theiia.org/about-us/about-ia/Pages/Position-Papers.aspx.

Jorion, P. (2007). *Value at risk*. New York: McGraw-Hill.

Kalberer, T., & Ravindran, K. (Eds.). (2009). *Variable annuities*. A global perspective: Risk Books.

Kirova, M., & Steinmann, L. (2012). Understanding profitability in life insurance. *SIGMA, (1/2012)*. https://media.swissre.com/documents/sigma1_2012_en.pdf.

Kirova, M., & Steinmann, L. (2013). Life insurance: Focusing on the consumer. *SIGMA, (6/2013)*. https://media.swissre.com/documents/sigma6_2013_en.pdf.

Klein, A. M. (2018). Underwriting around the World. Technical report, Mortality Working Group of the International Actuarial Association - Underwriting Subcommittee. https://www.actuaries.org/CTTEES_TFM/Documents/MWG_Underwriting_Report_May2018_AKlein.pdf.

Klugman, S. A., Panjer, H. H., & Willmot, G. E. (2012). *Loss models: From data to decisions*. Hoboken: Wiley.

Koller, G. (1999). *Risk assessment and decision making in business and industry. A practical guide*. Boca Raton: CRC Press.

Koller, M. (2011). *Life insurance risk management essentials. EAA series*. Berlin: Springer.

KPMG. (2001). *Enterprise risk management*. KPMG Australia: An emerging model for building shareholder value.

Lam, J. (2003). *Enterprise risk management. From incentives to controls*. Hoboken: Wiley.

Ledlie, M. C., Corry, D. P., Finkelstein, G. S., Ritchie, A. J., Su, K., & Wilson, D. C. E. (2008). Variable annuities. *British Actuarial Journal*, *14*(2), 327–389.

Mangiero, S. M. (2005). *Risk management for pensions. Endowment and foundations*. Hoboken: Wiley.

Martin, J., & Elliot, D. (1992). Creating an overall measure of severity of disability for the office of population censuses and surveys disability survey. *Journal of the Royal Statistical Society. Series A (Statistics in Society)*, *155*(1), 121–140.

Maurer, R., Rogalla, R., & Siegelin, I. (2013). Participating life annuities: Lessons from Germany. *ASTIN Bulletin*, *43*(2), 159–187.

McNeil, A. J., Frey, R., & Embrechts, P. (2005). *Quantitative risk management: Concepts, techniques, and tools*. Princeton: Princeton University Press.

Milevsky, M. A. (2005). Real longevity insurance with a deductible: Introduction to advanced-life delayed annuities (ALDA). *North American Actuarial Journal*, *9*(4), 109–122.

Milevsky, M. A. (2013). *Life annuities: An optimal product for retirement income*. Research Foundation of CFA Institute. http://www.cfapubs.org/toc/rf/2013/2013/1.

Müller, E., & Sandberg, D. (2020). Appropriate applications of stress and scenario testing. In *IAA Risk Book, chapter 19*. International Actuarial Association. https://www.actuaries. org/IAA/Documents/CMTE_IRC/RiskBook_project/Approved_Chapters/Ch19_Appropriate_ Applications_Stress_Testing_21Feb2020.pdf.

Niittuinperä, J. (2020). Policyholder behaviour and management actions. In *IAA Risk Book, chapter 18*. International Actuarial Association. https://www.actuaries.org/IAA/Documents/ CMTE_IRC/RiskBook_project/Revised_Chapters/Ch18_Policyholder_Behavior_20Feb2020. pdf.

O'Brien, C. (2013). Annuities: A complex market for consumers. Technical report. Nottingham University Business School. https://www.nottingham.ac.uk/business/businesscentres/crbfs/ documents/crbfs-reports/crbfs-paper5.pdf.

Olivieri, A., & Pitacco, E. (2002). Inference about mortality improvements in life annuity portfolios. In *Transactions of the 27th International Congress of Actuaries*, Cancun (Mexico). http://www.actuaries.org/EVENTS/Congresses/Cancun/ica2002_subject/mortality/ mortality_76_olivieri_pitacco.pdf.

Olivieri, A., & Pitacco, E. (2008). *L'Assurance-vie. Évaluer les Contrats et les Portefeuilles*. London: Pearson Education.

Olivieri, A., & Pitacco, E. (2009a). Solvency requirements for life annuities allowing for mortality risks: Internal models versus standard formulas. In M. Cruz (Ed.), *The Solvency II handbook. Developing ERM frameworks in insurance and reinsurance companies* (pp. 371–397). Risk Books.

Olivieri, A., & Pitacco, E. (2009b). Stochastic mortality: The impact on target capital. *ASTIN Bulletin, 39*(2), 541–563.

Olivieri, A., & Pitacco, E. (2012). Life tables in actuarial models: From the deterministic setting to a Bayesian framework. *AStA Advances in Statistical Analysis, 96*(2), 127–153.

Olivieri, A., & Pitacco, E. (2015). *Introduction to insurance mathematics. Technical and financial features of risk transfers. EAA series* (2nd ed.). Berlin: Springer.

Olivieri, A., & Pitacco, E. (2016). Frailty and risk classification for life annuity portfolios. *Risks, 4*(4), 39. http://www.mdpi.com/2227-9091/4/4/39.

Olivieri, A., & Pitacco, E. (2020). Linking annuity benefits to the longevity experience: Alternative solutions. *Annals of Actuarial Science*, 1–22.

Olson, D. L., & Wu, D. (Eds.). (2008). *New frontiers in enterprise risk management*. Berlin: Springer.

Orros, G., & Howell, J. (2008). Creating value through integrated ERM for healthcare insurers in Europe. In *ERM Symposium*, Chicago, Illinois. http://www.ermsymposium.org/2008/pdf/papers/ Orros.pdf.

Orros, G., & Smith, J. (2010). ERM for health insurance from an actuarial perspective. Sessional research paper, presented to the Institute and Faculty of Actuaries. http://www.actuaries.org.uk/ research-and-resources/documents/erm-health-insurance-actuarial-perspective.

Oxera Consulting. (2010). The use of gender in insurance pricing. Association of British Insurers, ABI research paper 24. www.eoc.org.uk/research.

Parmenter, D. (2015). *Key performance indicators. Developing, implementing and using winning KPIs* (3rd ed.). Hoboken: Wiley.

Pearson, N. D. (2002). *Risk budgeting*. Hoboken: Wiley.

Pitacco, E. (2007). Mortality and longevity: A risk management perspective. Invited lecture at the 1st IAA Life Section colloquium, Stockholm. http://www.actuaries.org/LIFE/Events/Stockholm/ Pitacco.pdf.

Pitacco, E. (2014). *Health insurance. Basic actuarial models. EAA series*. Berlin: Springer.

Pitacco, E. (2016a). Guarantee structures in life annuities: A comparative analysis. *The Geneva Papers on Risk and Insurance - Issues and Practice, 41*(1), 78–97.

Pitacco, E. (2016b). Premiums for long-term care insurance packages: Sensitivity with respect to biometric assumptions. *Risks, 4*(1), 1–22. http://www.mdpi.com/2227-9091/4/1/3.

Pitacco, E. (2017). Life annuities. Products, guarantees, basic actuarial models. CEPAR working paper 2017/6. http://cepar.edu.au/sites/default/files/Life_Annuities_Products_Guarantees_ Basic_Actuarial_Models_Revised.pdf.

Pitacco, E. (2019). Heterogeneity in mortality: A survey with an actuarial focus. *European Actuarial Journal, 9*(1), 3–30.

Pitacco, E., Denuit, M., Haberman, S., & Olivieri, A. (2009). *Modelling longevity dynamics for pensions and annuity business*. Oxford: Oxford University Press.

PwC. (2007). Guide to key performance indicators. Price water house Coopers. https://www.pwc.com/gx/en/audit-services/corporate-reporting/assets/pdfs/uk_kpi_guide.pdf.

Rejda, G. E. (2010). *Principles of risk management and insurance*. London: Pearson.

Reuss, A., Russ, J., & Wieland, J. (2015). Participating life insurance contracts under risk based solvency frameworks: How to increase capital efficiency by product design. In K. Glau, M. Scherer, & R. Zagst, (Eds.), *Innovations in quantitative risk management*. Springer proceedings in mathematics and statistics (Vol. 99, pp. 185–208). Berlin: Springer.

Reuss, A., Russ, J., & Wieland, J. (2016). Participating life insurance products with alternative guarantees: Reconciling policyholders' and insurer's interests. *Risks, 4*(11),

Rickayzen, B. D. (2007). An analysis of disability-linked annuities. Faculty of Actuarial Science and Insurance, Cass Business School, City University, London. Actuarial research paper no. 180. http://www.cass.city.ac.uk/__data/assets/pdf_file/0018/37170/180ARP.pdf.

Rickayzen, B. D., & Walsh, D. E. P. (2002). A multi-state model of disability for the United Kingdom: Implications for future need for long-term care for the elderly. *British Actuarial Journal, 8*(2), 341–393.

Ridsdale, B. (2012). Annuity underwriting in the United Kingdom. Note for the International Actuarial Association Mortality Working Group. http://www.actuaries.org/mortality/Item10_Annuity_underwriting.pdf.

Rinke, C. R. (2002). The variability of life reflected in annuity products. Issue no. 8: Hannover Re's Perspectives - Current Topics of International Life Insurance.

Rochette, M. (2009). From risk management to ERM. MPRA paper no. 32844. https://mpra.ub.uni-muenchen.de/32844/.

Rudelius, W., & Wood, G. L. (1970). Life insurance product innovations. *The Journal of Risk and Insurance, 37*(2), 169–184.

Rudolph, M. J. (2009). Enterprise Risk Management (ERM) practice as applied to health insurers, self-insured plans, and health finance professionals. Research report. Society of Actuaries, Health Section. Available at SSRN: http://www.soa.org/files/pdf/research-erm-pract-health.pdf.

Sandberg, D. (2015). Introduction to the IAA Risk Book. In *IAA Risk Book, chapter 1*. International Actuarial Association. https://www.actuaries.org/IAA/Documents/CMTE_IRC/RiskBook_project/Approved_Chapters/Ch19_Appropriate_Applications_Stress_Testing_21Feb2020.pdf.

SOA (2003). Improving the product development process. *Record of the Society of Actuaries, 29*(1). https://www.soa.org/globalassets/assets/library/proceedings/record-of-the-society-of-actuaries/2000-09/2003/may/rsa03v29n155ts.pdf.

SOA. (2017). Understanding the product development process of individual life insurance and annuity companies. Society of Actuaries. Available at SSRN: https://www.soa.org/resources/research-reports/2017/product-development-process/.

Sweeting, P. (2017). *Financial enterprise risk management*. Cambridge: Cambridge University Press.

Taylor, G. (2013). ERM in an optimal control framework. Research paper no. 2015ACTL17, Business School, UNSW, Sydney. Available at SSRN: https://ssrn.com/abstract=2660026.

Telford, P. G., Browne, B. A., Collinge, E. J., Fulcher, P., Johnson, B. E., Little, W., et al. (2011). Developments in the management of annuity business. *British Actuarial Journal, 16*(3), 471–551.

Tiller, J. E., & Fagerberg Tiller, D. (2015). *Life, health and annuity reinsurance* (3rd ed.). New Hartford: ACTEX Publications.

Williams, C. A., Smith, M. L., & Young, P. C. (1998). *Risk management and insurance*. New York: Irwin/McGraw-Hill.

Zhu, N., & Bauer, D. (2014). A cautionary note on natural hedging of longevity risk. *North American Actuarial Journal, 18*(1), 104–115.

Index

© Springer Nature Switzerland AG 2020
E. Pitacco, *ERM and QRM in Life Insurance*, Springer Actuarial,
https://doi.org/10.1007/978-3-030-49852-8

CPSIA information can be obtained
at www.ICGtesting.com
Printed in the USA
LVHW082159160222
711365LV00007B/410